Road to Net Zero

Oliver Zipse • Joachim Hornegger
Thomas Becker • Markus Beckmann
Michael Bengsch • Irene Feige
Markus Schober
Editors

Road to Net Zero

Strategic Pathways for Sustainability-Driven Business Transformation

Editors
Oliver Zipse
BMW AG
Munich, Germany

Thomas Becker
BMW AG
Munich, Germany

Michael Bengsch
BMW AG
Munich, Germany

Markus Schober
Friedrich-Alexander-Universität
Erlangen-Nürnberg
Erlangen, Germany

Joachim Hornegger
Friedrich-Alexander-Universität
Erlangen-Nürnberg
Erlangen, Germany

Markus Beckmann
Friedrich-Alexander-Universität
Erlangen-Nürnberg
Nuremberg, Germany

Irene Feige
BMW AG
Munich, Germany

ISBN 978-3-031-42223-2   ISBN 978-3-031-42224-9 (eBook)
https://doi.org/10.1007/978-3-031-42224-9

This is an Open Access publication.

Friedrich-Alexander-Universität Erlangen-Nürnberg

© The Editor(s) (if applicable) and The Author(s) 2023
Open Access This book is licensed under the terms of the Creative Commons Attribution 4.0 International License (http://creativecommons.org/licenses/by/4.0/), which permits use, sharing, adaptation, distribution and reproduction in any medium or format, as long as you give appropriate credit to the original author(s) and the source, provide a link to the Creative Commons license and indicate if changes were made.
The images or other third party material in this book are included in the book's Creative Commons license, unless indicated otherwise in a credit line to the material. If material is not included in the book's Creative Commons license and your intended use is not permitted by statutory regulation or exceeds the permitted use, you will need to obtain permission directly from the copyright holder.
The use of general descriptive names, registered names, trademarks, service marks, etc. in this publication does not imply, even in the absence of a specific statement, that such names are exempt from the relevant protective laws and regulations and therefore free for general use.
The publisher, the authors, and the editors are safe to assume that the advice and information in this book are believed to be true and accurate at the date of publication. Neither the publisher nor the authors or the editors give a warranty, expressed or implied, with respect to the material contained herein or for any errors or omissions that may have been made. The publisher remains neutral with regard to jurisdictional claims in published maps and institutional affiliations.

This Springer imprint is published by the registered company Springer Nature Switzerland AG
The registered company address is: Gewerbestrasse 11, 6330 Cham, Switzerland

Paper in this product is recyclable.

Foreword

When I was asked to write the foreword for this book, I hesitated. As the president of the Club of Rome, I am a staunch believer that the twenty-first century business, political and academic leadership must be much bolder and deeply systemic to face today's complex challenges.

As a BMW Sustainable Mobility Advisory Council member, I have consistently pushed the company to go further in its sustainability approach, not only as a vehicle manufacturer but also as a mobility provider, because I fundamentally believe that this is a company that can be a transformational mobility leader through its incredible engineering prowess and engrained social and environmental values.

To be honest, what has intrigued me most about this book is the process itself and the openness of the contributors in sharing their insights through a series of testimonials and interviews with BMW Chairman Oliver Zipse. This is a book that brings together the foundations for change to meet the twenty-first century needs for people, the planet and prosperity: dialogue, trust building, knowledge exchange, integrated non-linear thinking and acting. Room still remains for more expansive thinking on a complete shift in consumption patterns overall and on the move from circular to regenerative value chains; nevertheless, it is a sound exploration of net zero strategy building through an optimized industry–university partnership. Importantly, it is anchored in the most fundamental element of shifting from thought leadership to action by embracing co-design principles anchored in experimentation and learning, with clear examples of implementation practices, reporting and progress measurement.

There is no singular pathway to net zero and there most certainly is no silver technological bullet, especially without governance and economic transformation at the same time. But *Road to Net Zero: Strategic Pathways for Sustainability-Driven Business Transformation* serves as a window into how BMW has developed its sustainability strategy since 1973, when it became the first automotive manufacturer worldwide to add the position of Environmental Officer to its roster. This book is a practical guide full of lived experience and reflections on how to put into place a net-zero strategy whilst struggling as a company with complexity and twenty-first century pressure points.

What this book shows is that an automotive manufacturer and an innovative university can have a meaningful dialogue and set of programmes around industrial transformation and societal needs with a joint desire to address wicked problems and learn together as they move towards net-zero objectives.

In 1979, in his introduction to 'No Limits to Learning: Bridging the Human Gap', a report to the Club of Rome, Aurelio Peccei, Italian industrialist and founder of The Club of Rome, said, 'innovative learning is a necessary means of preparing individuals and societies to act in concert in new situations, especially those that have been and continue to be, created by humanity itself. Innovative learning, we shall argue, is an indispensable prerequisite to resolving any of the global issues'.

The exchanges and experiences across this book follow the tradition of innovative learning. This is a virtuous circle for change, where we take the time to write down what has worked and what hasn't, where we put in place exchanges and testimonials to ensure, as Aurelio Peccei says, that 'we learn what it takes to learn we should learn—and learn it'. Let us not forget that Peccei was an industrialist who understood the importance of innovation and deep systems change as fundamental pillars for human evolution. When Peccei spoke these words of wisdom almost 50 years ago, he already felt time was running out. That we needed to learn fast.

Today, we definitely do not have the luxury of time to get our net-zero journey wrong. 'No Limits to Learning' followed in the footsteps of the seminal report to the Club of Rome, 'The Limits to Growth', published in 1972 and showing that economic growth could not continue at the pace and scale predicted without pushing humanity beyond the planetary boundaries. That serious tipping points could occur in the 2020s. Here we are, fifty years later, amid a series of social and environmental tipping points and encapsulated in a poly crisis where the urge for knee-jerk short-term solutions abound, and yet we know that any decision made now has a multitude of serious long-term impacts. One example is shifting our dependency on gas from Russia to Africa or the Middle East due to the Ukrainian invasion rather than phasing out fossil energy and tripling investments in renewables and efficiency measures.

How can we ensure that our short-term firefighting reactions at all levels of society, from political to business decision making, build in long-term resilience to future shocks and stresses—not only to increasing climate impacts but also to wars, democratic and geopolitical instability, the migration of peoples and deep value chain disruptions? How do we all stay the course on our net-zero journey, when faced with so many short-term distractions?

Universities and industrial ecosystems have a fundamental role to play across today's societal fabric, serving as the light post for transformation, and we deeply need brave leadership and honest conversations to ensure that virtuous circle of change.

Now, of course, the challenge is to translate the wisdom across these pages—the 'learn it' part—into application, and that is my call to the authors of this publication and to Oliver Zipse and Joachim Hornegger as leaders in their own right. Use the depth of insight you have gained to transform not only your own institutions but also your own ecosystems and value chains globally.

Do stay on course. Address our global challenges head-on as the ultimate transformational experiment of our time. Let's shift our climate challenge from the greatest existential risk to humanity to the greatest opportunity for people, planet, and prosperity.

Sandrine Dixson-Declève

Co-President, The Club of Rome
Chair, the Economic and Societal Impacts of Research & Innovation Working Group, DG R&I, European Commission
Member, BMW Sustainable Mobility Advisory Council
Munich, Germany

Acknowledgements

On behalf of the editors, we sincerely appreciate the opportunity to unite diverse perspectives and inspiring facets of sustainability-driven business transformations in this edited volume. The work embodies a collaborative effort between academia and industry, illuminating the critical inquiries on the Road to Net Zero.

First and foremost, our profound thanks extend to all the key contributors to this edited volume. Your wealth of expertise and experience, sourced from both dedicated research and in-depth business practice, has greatly enriched the depth and breadth of this edited volume. Beyond synthesizing the relevant research aspects, your readiness to share your perspectives on sustainability-driven business transformation within each chapter has provided invaluable insights that will guide readers as they navigate the complexities of achieving a net-zero future.

We also express our heartfelt gratitude to the diligent individuals working behind the scenes—the internal and external reviewers and providers of feedback, proofreaders, designers, and the publishing team—who have committed their time, critical acumen, and attention to detail.

This edited volume wouldn't be as special without the support of Sandrine Dixson-Declève, Co-President of the Club of Rome, whose foreword sets the tone for this edited volume, stimulating readers with her thoughtful perspectives and underscoring the importance of collective action to achieve a net-zero future. We are privileged to have your wisdom, experience, and counsel on the first pages of this book.

Lastly, our thanks extend to you, the readers, and the broader sustainability community. This dynamic community, including its critical voices, consistently propels research and practice forward. We are grateful that you have joined us on this collective learning journey.

With deep gratitude,
The Editors
Oliver Zipse, Joachim Hornegger, Thomas Becker, Markus Beckmann, Michael Bengsch, Irene Feige and Markus Schober

About This Book

Sustainability is no longer simply a trend, customer preference, or political goal. It is the new benchmark aligning strategic objectives and measures in the automotive industry. To embark on a sustainable transformation, companies must adopt a science-based management approach, integrating various disciplines. This book is the culmination of insightful discussions among distinguished leaders from both academia and industry, collectively committed to driving the sustainability transformation of the automotive sector.

Recognizing that developing strategic pathways for sustainability-driven business transformation necessitates Pioneering Pathways, underpinned by strong university-industry partnerships (Chap. 1), every path towards achieving net-zero emissions commences with Setting the Course for Net Zero (Chap. 2) by translating climate science into actionable political and corporate targets.

The pursuit of ambitious climate targets propels us to the next stage—Crafting Corporate Sustainability Strategies (Chap. 3) and elaborating on The Future of Corporate Disclosure (Chap. 4). Subsequently, the transformational journey extends across the value chain, encompassing topics such as Creating Sustainable Products (Chap. 5), Transforming Value Chains for Sustainability (Chap. 6), and the pursuit of Sustainability in Manufacturing (Chap. 7). Along every path to net zero, The Power of Technological Innovation (Chap. 8) emerges, which, within the automotive industry, is defined, among other things, by the potential of novel drive systems. Finally, the end of each pathway marks the beginning of a new dawn, reflected in Chap. 9—The Road to Net Zero and Beyond.

By collating extensive thematic expert conversations and a comprehensive synthesis of research within each subject area, this book presents pivotal guiding questions that will drive the transformation towards sustainability. As an essential read for decision-makers, strategists, business developers, engaged citizens, and educators alike, this book offers valuable insights for navigating pathways towards a more sustainable future.

Contents

1 Pioneering Pathways   1
Joachim Hornegger and Oliver Zipse
- 1.1 A Collaborative Approach   1
- 1.2 Be Inspired   2
- 1.3 Expert Conversation on Joining Forces for Sustainability-Driven Transformation   4
- 1.4 Outline of the Book   12

2 Setting the Course for Net Zero   17
Markus Beckmann, Gregor Zöttl, Veronika Grimm, Thomas Becker, Markus Schober, and Oliver Zipse
- 2.1 Introduction   17
- 2.2 A Brief Review of Selected Climate Science Insights   18
- 2.3 Global Climate Policy: The Road to Paris and Beyond   22
- 2.4 The Role of National Policy Frameworks and Governance Mechanisms   26
  - 2.4.1 Market-Based Instruments to Directly Internalise External Costs   28
  - 2.4.2 Direct Support Instruments for Sustainable Technologies and Products   30
  - 2.4.3 Traditional Regulatory Approaches: 'Command-and-Control Measures'   32
- 2.5 Expert Conversation on Economic Climate Policy: Between Technological Openness and Regulation   34

	2.6	Science-Based Targets: Opportunities and Challenges of Setting Emissions Targets at the Company Level	43
		2.6.1 The Science-Based Target Initiative: Origin and Mission	43
		2.6.2 The Science Base of the SBTs	44
		2.6.3 Which Emissions Count? Clarifying the Scope and Base Year for SBTs	45
		2.6.4 Different Types of Science-Based Targets	46
		2.6.5 Different Methods and Sector Approaches for Determining Necessary Reduction Levels	47
		2.6.6 Benefits and Challenges of Science-Based Targets	49
	2.7	Conclusion	51
	References		53

3 Crafting Corporate Sustainability Strategy 61
Markus Beckmann, Thomas Becker, and Oliver Zipse

3.1	Introduction	61
3.2	Strategy Development and Sustainability: Past and Present	62
3.3	Expert Conversation on Integrating Sustainability into Corporate Strategy	70
3.4	The Future of Integrated Strategies: Challenges, Opportunities, and Key Questions	82
3.5	Conclusion	86
References		87

4 The Future of Corporate Disclosure 93
Thomas Fischer, Jennifer Adolph, Markus Schober, Jonathan Townend, and Oliver Zipse

4.1	Introduction	93
4.2	New Ways of Reporting	95
4.3	The Current Legislative Landscape in the EU	98
4.4	Integrated Reporting in Practice	102
4.5	Expert Conversation on the Implementation of Integrated Reporting at BMW Group	103
4.6	The Future of Reporting: Opportunities, Challenges and the Role of Integrated Reporting	109
	4.6.1 Opportunities of New Ways of Reporting	109
	4.6.2 Challenges of New Ways of Reporting	111
	4.6.3 The Future Potential of Integrated Reporting	113

		4.7	Conclusion	114
			References	115

5 Creating Sustainable Products — 123
Lena Ries, Sandro Wartzack, and Oliver Zipse

- 5.1 Introduction — 123
- 5.2 Pathways Toward Circular Design — 124
  - 5.2.1 The CE Approach Offers a Paradigm Shift — 126
  - 5.2.2 Different Frameworks for CE Operationalization: Slowing, Closing, Narrowing, and R-Strategies — 127
  - 5.2.3 Three Implications for Design — 128
    - 5.2.3.1 First Implication: A Change in Product Design — 129
    - 5.2.3.2 Second Implication: A Change in Service Design — 130
    - 5.2.3.3 A Third Implication: A Change in User Behavior — 132
  - 5.2.4 Implementation Challenges — 133
- 5.3 Expert Conversation on Sustainability in Product Development — 134
- 5.4 The Future of Sustainable Product Development — 143
  - 5.4.1 Digital Technologies as Enablers of CE — 143
  - 5.4.2 Better Together: The Need for Broadening Perspectives — 145
- 5.5 Conclusion — 147
- References — 148

6 Transforming Value Chains for Sustainability — 159
Kai-Ingo Voigt, Lothar Czaja, and Oliver Zipse

- 6.1 Introduction — 159
- 6.2 In the Age of Resource Scarcity — 160
- 6.3 Value Chain Transformations — 163
  - 6.3.1 Recycling of Lithium-Ion Batteries (LIBs) — 165
  - 6.3.2 Reverse Logistics (Phase 1) — 166
  - 6.3.3 Pretreatment (Phase 2) — 167
  - 6.3.4 Metallurgical Treatment (Phase 3) — 168
  - 6.3.5 Reintroduction into the Market (Phase 4) — 170
- 6.4 Expert Conversation on Sustainability in the Supply Chain — 171

	6.5	Closing the Loop	176
	6.6	Outlook and Further Challenges	179
	6.7	Conclusion	182
	References		183

7 Sustainability in Manufacturing Transforming — 187
Nico Hanenkamp and Oliver Zipse

- 7.1 Introduction — 187
- 7.2 The Three Dimensions of Sustainable Production — 188
  - 7.2.1 Practical Perspectives on Sustainable Manufacturing — 190
  - 7.2.2 Research Perspectives on Sustainable Manufacturing — 191
- 7.3 Expert Conversation on Sustainability in Production — 195
- 7.4 The Sustainable Factory of the Future — 203
  - 7.4.1 Sustainable Manufacturing Processes along the Life Cycle — 204
  - 7.4.2 Digitalization, Artificial Intelligence, and IoT — 206
  - 7.4.3 Application of Sustainability Standards — 207
  - 7.4.4 System Coupling, Urban Production, and Smart Cities — 208
  - 7.4.5 Summary and Outlook — 209
- 7.5 Conclusion — 210
- References — 211

8 The Power of Technological Innovation — 215
Jörg Franke, Peter Wasserscheid, Thorsten Ihne, Peter Lamp, Jürgen Guldner, and Oliver Zipse

- 8.1 Introduction — 215
- 8.2 An Overview on Alternative Drive Systems — 216
- 8.3 Expert Conversation on the Future of Mobility — 226
- 8.4 Expert Conversation on H_2 as Fuel of the Future — 235
- 8.5 Beyond Technical Functionality: The Energy Ecosystem around Eco-Efficient Drivetrain Solutions — 245
  - 8.5.1 Eco-Efficient Storage of Electric Energy on Board an Automobile — 245
  - 8.5.2 Eco-Efficient Hydrogen Storage and On-Demand Electrification — 251
  - 8.5.3 Eco-Efficient Electricity Distribution via an Electrified Road Infrastructure — 255
- 8.6 Conclusion — 257
- References — 260

9	**The Road to Net Zero and Beyond**		**265**
	Markus Beckmann and Irene Feige		
	9.1	Reflecting on Collaborative Learning	265
	9.2	A Summary of this Book's Storyline	266
		9.2.1 Chapter 2: Setting the Course for Net Zero	266
		9.2.2 Chapter 3: Crafting Corporate Sustainability Strategy	267
		9.2.3 Chapter 4: The Future of Corporate Disclosure	267
		9.2.4 Chapter 5: Creating Sustainable Products	268
		9.2.5 Chapter 6: Transforming Value Chains for Sustainability	268
		9.2.6 Chapter 7: Sustainability in Manufacturing	269
		9.2.7 Chapter 8: The Power of Technological Innovation	270
		9.2.8 Looking Back on the Road to Net Zero	270
	9.3	Insights and Themes across Chapters	271
		9.3.1 Sustainability Demands Both Integrative Thinking and Integrative Management	271
		9.3.2 Sustainability Is a Moving Target	272
		9.3.3 Sustainability Is a Race You Cannot Win Alone	273
		9.3.4 It Is All About Data: Measurable Indicators, Targets, Transparency, and Digitalization	275
		9.3.5 Not Everything That Matters Can Be Measured (Accurately)	277
	9.4	Beyond the Road to Net Zero	278
		9.4.1 Beyond Decarbonization	278
		9.4.2 Beyond Reducing Negative Impacts	280
		9.4.3 Beyond Single Trajectories That Ignore the Role of Space	282
	9.5	The Future of Industry–University Partnerships	283

About the Editors and Contributors

About the Editors

Prof. Dipl.-Ing. Oliver Zipse is Chairman of the Board of Management of BMW AG since 2019. He joined BMW AG in 1991 and has held various responsibilities within the company in development, technical planning and production in Munich, South Africa and the UK. He holds a Dipl. Ing. Degree from the Technical University of Darmstadt and an Executive MBA from the WHU Koblenz and Kellogg School of Management Evanston/USA. In addition, he is an honorary professor at the Technical University of Munich.

Prof. Dr.-Ing. Joachim Hornegger is President of the Friedrich-Alexander-Universität Erlangen-Nürnberg (FAU), Germany and former Chair of Pattern Recognition at the Faculty of Engineering (FAU). During his professional career, he was a guest researcher at the Massachusetts Institute of Technology (MIT) and the Computer Science Department at Stanford University. In addition to his research activities, he held various positions in the medical technology industry at Siemens Medical Solutions.

Dr. Thomas Becker is Vice President of Sustainability and Mobility at BMW AG, Germany. After studying business administration at the University of Cologne (Germany), he completed his doctorate at the University of Witten/Herdecke (Germany) before working as an environmental policy officer at the Federation of German Industries (BDI) and later as Deputy Managing Director of the German Association of the Automotive Industry (VDA).

Prof. Dr. Markus Beckmann holds the Chair for Corporate Sustainability Management at FAU Erlangen-Nürnberg, Germany. Prior to his appointment at

FAU, he worked as an assistant professor of social entrepreneurship at the Centre for Sustainability Management, Leuphana University of Lüneburg (Germany). His research focuses on sustainability management and social entrepreneurship, as well as business ethics and corporate social responsibility.

Michael Bengsch works in the Corporate Strategy Department for Sustainability and Mobility at BMW AG, Germany. He studied electrical and computer engineering at the Technical University of Munich (Germany).

Dr. Irene Feige is Head of Climate Strategy and Circular Economy at BMW AG, Germany. After studying economics at Vienna University of Economics and Business, she completed her doctorate in economics at the University of Innsbruck, while also taking part in research stays in Cambridge (US), Beijing and Shanghai. During her career at BMW she has taken on several positions in corporate strategy, communication and research.

Markus Schober works in strategy consulting for the automotive industry and is a start-up entrepreneur. Previously, he worked for several years as university-industry relations manager at FAU Erlangen-Nürnberg, Germany. He studied Industrial Engineering & Management at FAU Erlangen-Nürnberg and East China University of Science and Technology, Shanghai.

Contributors

Jennifer Adolph, M.Sc. Research Associate at the Chair for Corporate Sustainability Management at Friedrich-Alexander-Universität Erlangen-Nürnberg, Nuremberg, Germany

Thomas Becker, Dr. Vice President Sustainability and Mobility at BMW AG, Munich, Germany

Markus Beckmann, Prof. Dr. Head of the Chair for Corporate Sustainability Management (Full Professor) at Friedrich-Alexander-Universität Erlangen-Nürnberg, Nuremberg, Germany

Lothar Czaja, Dr. Postdoc at the Chair of Industrial Management at Friedrich-Alexander-Universität Erlangen-Nürnberg, Nuremberg, Germany

Irene Feige, Dr. Head of Climate Strategy and Circular Economy at BMW AG, Munich, Germany

Thomas M. Fischer, Prof. Dr. Head of the Chair for Accounting and Management Control (Full Professor) at Friedrich-Alexander-Universität Erlangen-Nürnberg, Nuremberg, Germany

Jörg Franke, Prof. Dr-Ing. Head of the Institute for Factory Automation and Production Systems (Full Professor) at Friedrich-Alexander-Universität Erlangen-Nürnberg, Erlangen, Germany

Veronika Grimm, Prof. Dr. Head of the Chair of Economic Theory (Full Professor) at Friedrich-Alexander-Universität Erlangen-Nürnberg, Nuremberg, Germany

Jürgen Guldner, Dr. General Program Manager Hydrogen Technology at BMW AG, Munich, Germany

Nico Hanenkamp, Prof. Dr.-Ing. Head of the Institute of Resource and Energy Efficient Production Machines (Full Professor) at Friedrich-Alexander-Universität Erlangen-Nürnberg, Fuerth, Germany

Joachim Hornegger, Prof. Dr.-Ing. President of Friedrich-Alexander-Universität Erlangen-Nürnberg, Erlangen, Germany

Thorsten Ihne, M.Sc. Research Associate at the Institute for Factory Automation and Production Systems at Friedrich-Alexander-Universität Erlangen-Nürnberg, Erlangen, Germany

Peter Lamp, Dr. General Manager Battery Cell Technology and Fuel Cell at BMW AG, Munich, Germany

Lena Ries, M.A. Research Associate at the Chair for Corporate Sustainability Management at Friedrich-Alexander-Universität Erlangen-Nürnberg, Nuremberg, Germany

Markus Schober, M.Sc. University-Industry Relations Manager at Friedrich-Alexander-Universität Erlangen-Nürnberg, Erlangen, Germany

Jonathan Townend Head of Group Accounting & Reporting & Taxes at BMW AG, Munich, Germany

Kai-Ingo Voigt, Prof. Dr. Head of the Chair of Industrial Management (Full Professor) at Friedrich-Alexander-Universität Erlangen-Nürnberg, Nuremberg, Germany

Sandro Wartzack, Prof. Dr-Ing. Head of the Chair of Engineering Design at Friedrich-Alexander-Universität Erlangen-Nürnberg, Erlangen, Germany

Peter Wasserscheid, Prof. Dr. Head of the Chair of Chemical Engineering I – Reaction Engineering (Full Professor) at Friedrich-Alexander-Universität Erlangen-Nürnberg, Erlangen, Germany

Oliver Zipse, Prof. Dipl.-Ing. Chairman of the Board of Management of BMW AG, Munich, Germany

Gregor Zöttl, Prof. Dr. Holder of the Professorship of Economics (Industrial Organization and Energy Markets) at Friedrich-Alexander-Universität Erlangen-Nürnberg, Nuremberg, Germany

List of Abbreviations

A4S	Accounting for Sustainability
AC	Alternating current
ACA	Absolute contraction approach
ACES	Autonomous, connected, electric, shared
AI	Artificial Intelligence
AM	Additive manufacturing
APQP	Advanced product quality processes
ASSB	All-solid-state-battery
BCCC	Battery Cell Competence Centre
BEV	Battery electric vehicles
BMW	Bayerische Motoren Werke
C2C	Cradle-to-cradle
CDM	Clean development mechanism
CDP	Carbon Disclosure Project
CE	Circular economy
CMCC	Cell Manufacturing Competence Centre
CO_2	Carbon dioxide
COP	Conference of the Parties
CSRD	Corporate Sustainability Disclosure Directive
DC	Direct current
DoE	Design of experiments
EBT	Earnings before tax
EEG	Erneuerbare-Energien-Gesetz
EFRAG	European Financial Reporting Advisory Group
ELV	End-of-life vehicle
EMAS	Eco Management and Audit Scheme
EOL	End of life
ESRS	European Sustainability Reporting Standards

ESS	Energy storage systems
ETS	Emissions trading system
EU ETS	European Union Emissions Trading System
EU	European Union
EV	Electric vehicle
FAU	Friedrich-Alexander-Universität Erlangen-Nürnberg
FCEV	Fuel cell electric vehicles
FHEV	Full hybrid electric vehicles
FLAG	Food, land use and agriculture sector
GHG	Greenhouse gas
GRI	Global Reporting Initiative
GWh	Gigawatt hour
H_2	Hydrogen
HEAT	Highly-integrated electric drive train
HI ERN	The Helmholtz Institute Erlangen-Nürnberg for Renewable Energy
HVAC	Heating, ventilation, and air conditioning systems
ICE	Internal combustion engine
IEA	International Energy Agency
IFRS	International Financial Reporting Standards
IIRC	International Integrated Reporting Council
IoT	Internet of Things
IPCC	Intergovernmental Panel on Climate Change
IPCEI	Important Project of Common European Interest
IPT	Inductive power transfer
IR	Integrated Reporting
IRF	Integrated Reporting Framework
ISO	International Standardization Organization
ISSB	International Sustainability Standards Board
JI	Joint implementation
KPI	Key performance indicator
LCA	Life cycle assessment
LCO	Lithium-Cobalt-Oxide
LFP	Lithium ferro phosphate
LIB	Lithium-ion battery
LMO	Lithium-manganese oxide
LOHC	Liquid organic hydrogen carrier
MCHEV	Micro hybrid electric vehicle
MCI	Material Circularity Indicator
MHEV	Mild hybrid electric vehicle
NDCs	Nationally determined contributions
NGO	Non-governmental organization
NMC	Lithium nickel manganese cobalt oxide
OEE	Overall equipment effectiveness

OEM	Original equipment manufacturer
PHEV	Plug-in hybrid electric vehicles
PSS	Product-service system
R&D	Research and development
SBT	Science-based target
SBTi	Science-Based Targets initiative
SBTN	Science-based targets for nature
SCM	Supply chain management
SDA	Sectoral decarbonization approach
SDGs	Sustainable development goals
SME	Small and medium-sized enterprise
SoT	State of technology
TWh	Terawatt hour
UN	United Nations
UNEP	United Nations Environmental Programme
UNFCCC	United Nations Framework Convention on Climate Change
VUCA	Volatile, uncertain, complex, ambiguous
WBCSD	World Business Council for Sustainable Development
WLTP	Worldwide harmonized Light Vehicles Test Procedure
WMO	World Meteorological Organization
WRI	World Resources Institute

List of Figures

Fig. 1.1	Outline of the Road to Net Zero	13
Fig. 4.1	Elements in an integrated report to explain a firm's value creation, preservation or erosion as applied by the BMW Group	98
Fig. 5.1	R-strategies increasing circularity	128
Fig. 8.1	Overview of the competitive environment in the automotive industry	217
Fig. 8.2	Push and pull factors underpinning the success of electromobility	218
Fig. 8.3	Functionalities of different hybridization strategies	220
Fig. 8.4	Overview of various forms of function integration of electric motors	221
Fig. 8.5	Series drivetrains based on different types of electric motors	223
Fig. 8.6	Powertrain of the BMW iX5 Hydrogen with fuel cell stack, electric motor, and two high-pressure hydrogen tanks	224
Fig. 8.7	The efficiency of different powertrain options (*BEV* Battery Electric Vehicle, *SoT* State of Technology, *FCEV* Fuel Cell Electric Vehicle)	224
Fig. 8.8	Ragone diagram showing specific energy and power densities of different battery technologies	246
Fig. 8.9	Overview of the working principle of a Li-ion battery cell, the typical set of materials and the typical realization from materials to electrodes and finally to cells (pouch, cylindrical, or prismatic hard case)	247
Fig. 8.10	BMW's Gen5 battery cell, module, and pack architecture	248
Fig. 8.11	BMW's Gen6 battery cell and pack architecture and the resulting improvements of some relevant KPIs (the reference is the present Gen5 battery pack architecture)	249

Fig. 8.12	BMW's circular battery value chain	250
Fig. 8.13	Future diversified development directions for battery cell technology	251
Fig. 8.14	Illustration of the working principle of chemical hydrogen storage	254
Fig. 8.15	Inductive charging is based on resonant magnetic flux coupling	256

1

Pioneering Pathways

Universities and Industry as Collaborative Learners on the Road to Net Zero

Joachim Hornegger and Oliver Zipse

1.1 A Collaborative Approach

This book is the outcome of a joint experiment—an experimental exercise in university–industry relationship building between our institutions: BMW and Friedrich-Alexander-Universität (FAU). To be sure, our institutions have always had good and trusting relationships between individual experts—long before our time in leadership roles. People at BMW and FAU have been involved in joint research projects, joint student mentoring, guest lectures, expert advice, talent exchanges, and more. In fact, we admire these multifaceted forms of collaboration based on individual expertise or serendipity. But we felt there could and should be more.

Could we somehow extend our collaborative relationships to jointly address aspects of one of the dominant challenges of our time, namely the sustainable transformation of organisations? How could we make this happen? Could we create an inspiring and inclusive journey for many? We wanted to start small but with a sustainable perspective. We wanted to start informally but with measurable results. We wanted to use pragmatic approaches with scientific rigour and practical relevance. Above all, however, we wanted to invite others

---

J. Hornegger (✉)
FAU Erlangen-Nürnberg, Erlangen, Germany
e-mail: praesident@fau.de

O. Zipse
BMW AG, Munich, Germany

to join us on our journey on 'The Road to Net Zero'.[1] The time to do so is now. So feel free to aim even higher: Be inspired.

1.2 Be Inspired

Our ideas grew over a business lunch. Wouldn't it be great, we wondered, if our teams could be part of our conversation, join the journey, and experience the same inspiration and learning that we had experienced at our first lunch? As a first step, we invited a range of experts from across our organisations to participate in a series of conversations on topics involving sustainability-driven business transformation, with Mission Net Zero in mind. It was inspiring to see the level of interest and engagement. Despite the pandemic, a total of nine conversations took place over the course of 2021. From FAU:

- Prof. Dr.-Ing. Joachim Hornegger, President of FAU Erlangen-Nürnberg, opened the dialogue series with a discussion on the role of collaboration between academia and industry in addressing global sustainability challenges
- Prof. Dr. Veronika Grimm, Head of the Chair of Economic Theory at FAU and member of the German Council of Economic Experts, engaged in a conversation on climate policies needed to meet the objectives of the Paris Agreement
- Prof. Dr. Markus Beckmann, Head of the Chair for Corporate Sustainability Management, started a conversation on sustainability in corporate strategies
- Prof. Dr. Thomas Fischer, Head of the Chair of Accounting and Management Control, focused on integrated sustainability reporting
- Prof. Dr.-Ing. Sandro Wartzack, Head of the Institute of Engineering Design, looked at sustainability in product development
- Prof. Dr. Kai-Ingo Voigt, Head of the Chair of Industrial Management, focused on sustainability in the supply chain
- Prof. Dr.-Ing. Nico Hanenkamp, Head of the Institute of Resource and Energy Efficient Production Machines, engaged in a conversation on sustainability in production
- Prof. Dr.-Ing. Joerg Franke, Head of the Institute of Factory Automation and Production Systems, focused on the future of e-mobility and was joined by
- Prof. Dr. Peter Wasserscheid, Head of the Institute of Chemical Reaction Engineering and Director of the Helmholtz Institute Erlangen-Nürnberg, shared his expertise in hydrogen (H2)—the fuel of the future

---

[1] In the remainder of this book, we employ capital letters when using the phrase "Road to Net Zero" to denote our understanding of sustainability-driven business transformation.

The BMW Group was represented by Prof. Oliver Zipse, Chairman of the Board of Management of BMW AG, who was joined by:

- Dr. Thomas Becker, VP of Sustainability and Mobility
- Jonathan Townend, Head of Group Accounting & Reporting & Taxes
- Dr. Peter Lamp, General Program Manager Battery Cell Technology & Fuel Cell
- Dr. Jürgen Guldner, General Program Manager Hydrogen Technology

We all truly enjoyed the discussions, had them recorded, and initially published some short excerpts on the BMW website.[2] The feedback motivated us to go one step further. We decided to publish them, with some additional context, to inspire others to join our journey and do better, dig deeper, engage and elaborate further.

This book does exactly that. It will not pave the Road to Net Zero. Instead, it will show you that your contributions, deep conversations, collaborative work, and commitment are needed to take the necessary steps. This book is not written as a blueprint solution to the challenge of our time (in fact, you may feel that it often only scratches the surface), but it aims to outline a Road to Net Zero and create collective engagement. This book is not a political statement from BMW or FAU. It is what it is: a collection of edited expert conversations, framed by some foundational thoughts and complemented by our ideas about the next steps. It can serve as a primer for shaping industry–university relationships, starting with matchmaking (or should we say 'speed dating') of experts from both worlds—academia and industry—driven by a common mission and committed to bridging their individual spheres of knowledge. Conversations can help build trusting relationships; they are the gateway to deeper discussions, constructive mutual criticism, promising projects, and joint steps that pave the Road to Net Zero.

Our shared concern about climate change was the catalyst for our conversation. Now, we would like to invite you, the reader, to listen to our first exchange. The dialogue is about getting to know each other, exploring our positions and commitments, finding our common understanding and working out the complementarities of our institutions as a first step. You will find that we are beginning to feel our way along a road that was previously only imagined—an experience that has been equally fascinating, at times eye-opening and encouraging. We have identified a shared passion for open challenge-based collaboration, a shared interest in the promise and pitfalls of measurement (how can science help?), and confidence in the ability of our

---

[2] https://www.presstopic.bmwgroup.com/en/sustainability-dialogues

organisations to work fruitfully together on the sustainability challenge. This is where we started.

1.3 Expert Conversation on Joining Forces for Sustainability-Driven Transformation

Hornegger: Climate change is a serious issue that affects the lives of so many people. We simply have to address it to ensure that life on Earth, as we know it today, can continue. We all have a responsibility, especially those who lead a company, a university, or a larger group of people. FAU and BMW, the BMW Group, are certainly leaders in innovation—as well as in terms of climate protection, in terms of sustainability. BMW plays a sort of pioneering role in the automotive industry. What I would like to know, Mr. Zipse: At what point did you realise that this was a huge opportunity and, at the same time, a very demanding challenge?

Zipse: Professor Hornegger, thank you for the opportunity to speak to you today. Sustainability has been part of BMW's DNA for many, many years. We had our first sustainability manager as early as back in the 1970s. At that time, the focus was on the impact of production on the environment. Then, in 2008/2009, we conceived a new product. It was the i3, and it was designed to be fully sustainable. That's when the next step began: to put sustainability at the heart of BMW's strategy. Now, we have taken the next step to underline our pioneering role. Sustainability is not just a 'product thing'; it is at the core of our company strategy.

Hornegger: What is driving this development?

Zipse: Climate change is one factor, but sustainability is much more than climate change. There are several reasons for this: First, all resources are finite. So, if you are a major car manufacturer like us, you have to create a broader awareness, and you have to manage the resources. In addition, today, everything can be measured—through very cheap sensors, through the digitisation of our world. And when it's measured, it's transparent. Sustainability has a lot to do with transparency. These are some reasons why we decided to take the next step in our sustainability strategy.

Sustainability is an issue that affects every part of society. On a global scale, it affects every country, and science and scientific progress have a crucial role to play in understanding what it really means, in all its implications, in all its systemic features. What kind of role can science—and FAU in particular—play in making progress, in understanding what sustainability really means?

Hornegger: This is a very interesting question, because sustainability is not a closed field of research that we would deal with in a single department. It is an issue that cuts across the whole university, and it is present at different levels. If we look at basic research, we are trying to understand the fundamental mechanisms of nature. They have huge implications for sustainability. For example, how do you convert light energy into electrical energy?

Zipse: Understanding this question is very important for the renewable energies of the future.

Hornegger: Indeed. But merely understanding the basics is not enough. The second level is to use the knowledge gained from basic research to design new technologies and develop new approaches to solving practical problems. This is where we build systems; we develop new technologies, for example, to store hydrogen using chemical mechanisms. The third level, where we are very active, is closer to application. Here, too, we take an engineering approach: How can the technologies that work be scaled up for industrial use? Finally, the fourth level is equally important and characteristic of FAU: deep reflection. Here, we ask even broader questions: What is the economic impact? What is the impact on our society? How does it change the way we live and the way we interact with each other? In our FAU system, we have strong competencies at these different levels. I think we cover a very broad and well-connected spectrum where we can support our industrial partners with the needs they define and the problems they face.

Zipse: Would you underline a statement where we can say 'one plus one equals three' when we combine our activities?

Hornegger: Of course. I would even exaggerate a little and say that in this case, 'one plus one equals eleven' [laughter]. Just look at the strength of German engineering in industry and the strength of our university. It is not as if we at the FAU sit on an island and solve individual problems without looking to the right or left. We look at these problems in context; we think in terms of systems—we build systems. We also analyse the working systems and their effects. This holistic approach is the strength of our university—and is valued by many of our industry partners. It is also the basis of our innovation power.

Zipse: Well, and for Germany—and Bavaria in particular—to be a world leader in innovation, in industrial terms—you also mentioned the entire automotive industry, not only one car manufacturer here. I think there is a unique opportunity. Especially when we think about very complex systems, how to design them and how to understand them . . . science can play a big role as well in system integration methods.

Hornegger: That is also our experience. I think this is a clear advantage of our German engineering education system.

Zipse: So, let's use it!

Hornegger: We will use it! Let us talk about how you have integrated sustainability into your company strategy. I can imagine that you faced a lot of challenges when you started discussing this. What was the biggest issue in your company regarding sustainability as part of BMW's strategy?

Zipse: I can still remember the early days. I have been with the company for 30 years, and about 20 years ago, there was a discussion about how to combine sustainability with sporty cars. At first, it seemed like a contradiction. We then coined the term 'Efficient Dynamics'. And look: We had cars that were very dynamic, but at the same time—compared to our industry competitors—the least polluting ones. There was indeed a way to combine efficiency with dynamics in our cars.

Hornegger: So, while others saw efficiency and dynamics as a trade-off, you squared the circle?

Zipse: I am convinced that it is possible to combine competing objectives into one strategy. I think it is possible to be a major industry player and, at the same time, make a significant contribution to sustainability. This has a lot to do with being a pioneer, and the biggest hurdle is articulating it. Recognising that it is important is one thing, but expressing it, formulating a strategy and putting it into your strategy process—I think that is the biggest hurdle. If there is real action behind it, once you have communicated it, you will get a lot of replication. People start talking about it, and they start multiplying it. This is a task for every manager: not only to understand, but also to start communicating. That is, by the way, why we're sitting here.

Hornegger: Absolutely. Communication is key. You are relying on all of your 120,000 employees to buy into this concept. So, the transformation process is a very, very long-term transformation. How do you guide your employees through this process? How do you make them aware that this is one of the key strategic objectives you have defined?

Zipse: Well, I think by saying that this is the cornerstone of our strategy, and that it is not a contradiction: to build the best cars in the world, which are not only dynamic but also excellent to drive, and which have very low pollution impact. All our people are part of this new strategy. If you want them to be part of a new strategy, you need a lot of supporters. That, again, of course, means you have to talk about it and discuss it together. You have to explain it, of course. You cannot just propagate it. You have to implement through action, through setting targets.

Hornegger: Why is setting goals so important?

Zipse: I always say, 'What gets measured, gets done'. Setting a clear target for 2030, not just 2050, is very important. It is not difficult to say that we will be carbon neutral by 2050, according to the Paris Climate Agreement. Of course, we will, but that is so far away. It is almost 30 years away. That is why I want to hold us accountable, to make progress that starts today. We have set targets for 2030. We have also made clear that it is not just about the emissions from our cars. It is also about our supply chain, our administrative processes, our contracts with suppliers, our own production. Our targets are obviously about the whole life cycle of the car. We take that fully into account. 'Well to wheel' is our strategy, not just the car itself.

Hornegger: Let us also talk a little bit about the link to research, to basic research and to universities. When we talk about sustainability and the transformation process, many, many questions still need to be answered by research, and universities can also contribute to the progress in this area. The 'U' in university somehow stands for 'understanding'. FAU, as a whole, is a full-spectrum university, covering a wide range of different disciplines. They all contribute with their perspectives. Perhaps you could comment a little on that. What are your expectations regarding the cooperation between universities and industry?

Zipse: I think it is essential—not only for educational reasons—to build a bridge between science, education, the academic world, and management. Management today is becoming more and more science-based. Look at the coronavirus. In the search for a remedy, a science-based approach was taken. Good management is always linked to sound knowledge. Where does knowledge come from? A lot of it comes from universities—not just at the laboratory level, but also from systems thinking: How do whole systems work? What does sustainability mean for financial reporting? What does sustainability mean for education? How much do you need to spend on R&D to find the right solutions? I think there is a very close link and a lot of overlap between the scientific world and management.

Hornegger: How do you make this overlap work?

Zipse: I like the idea of science-based management. Our strategy is very much linked to an initiative called 'Science-Based Targets'. Why is this important? Because today, you can measure almost anything. You can immediately correlate the effects of your management decisions with facts. So, good solutions are always measured against other good solutions based on evidence. That is why I am working—also as a member of the Fraunhofer Society—to build a bridge to science. Of course, this day here at the FAU is also very symbolic of building this bridge because being successful in science and being successful as an organisation are highly correlated.

Hornegger: I totally agree with you, Mr. Zipse. When you look at our university, there is always the question of how we define its goals. Do we want to look at research output? Do we want to look at the educational programmes? Do we look at the appreciation of our educational programme among students? These are typical measures that we accept to define our future goals. The fact that BMW and the BMW Group have made sustainability a strategic cornerstone is something that I will now take with me. I will also initiate deeper discussions within our system: What is the goal of our university in terms of sustainability? How do we measure it?

Zipse: Do you see a similar value, as we do, in setting targets and measuring progress?

Hornegger: I like your statement that 'what you can measure really moves forward'. Well, in my experience, that is an insight I can support. If you can quantify things, you can show how they are changing and whether they are changing in the right direction. To be fair, we in universities are not very used to measuring performance and analysing whether we are on the right track. Here, I think we have a lot to learn from industry in terms of sustainability. In fact, our students are very motivated; our students are working hard on sustainability issues. They have developed a climate concept for our university. They presented it to the University Council, and they encouraged us to implement it at our university. They even pushed us to establish a Green Office, where we are looking at the goals they have set and what we can do next.

Zipse: This bottom-up student support for sustainability is very valuable.

Hornegger: Yes, and these steps are important to raise awareness in a system like FAU, where we also have about fifteen thousand employees. We are the second-largest employer in the region. In the cities of Nürnberg, Erlangen, and Fürth, we have a real responsibility. I appreciate this kind of input from you. I also think that sustainability can guide us in research. For example, our students are looking at issues such as electromobility or the challenges of hydrogen technology and how it can be applied to future drivetrains and energy systems. What approaches do you see working with us on the different levels, from basic research to the engineering, application, and reflection levels?

Zipse: Can I first make a brief statement about 'you don't measure at universities'? You grade every day. It is the toughest measurement you can imagine. Universities are actually used to measuring because they grade their students. I can still remember that grading was the biggest hurdle, you know, that you have to get good grades at the end …

Hornegger: Interesting. I never noticed that connection. Thanks for that comment!

Zipse: So, you are actually used to that. Now, let me try to answer about the similarities we have. Forty, fifty thousand students and staff together. That is a very large organisation. BMW has more than 150,000 employees worldwide on a global scale. Overall, the question is the same: How do you organise very large organisations? How do you set goals? Realistic targets? How do you manage progress? How do you manage change? How do you manage transformation? There are a few rules you need to follow to be successful.

Hornegger: I am all ears. What are these rules?

Zipse: The first thing is inclusion. You need a platform where all opinions can be expressed in order to find the best solution—especially at the university, where you have a lot of diverse knowledge. You have to have a platform where you—where everyone—can speak up and be part of a transformation process. The days of a small group—or, even worse, one person—deciding where to go are over. You can try that, but your progress will be very slow. So, I think inclusion is critical. Then, of course, how do you acquire new knowledge? How do you gather facts? You have to have a process for making decisions. You have to have a forum; you need meetings. Meeting management sounds very boring, but it is essential. How do you organise decision-making? If you don't decide, you can make a lot of speeches and say, 'We are transforming'. The question is: How and where do we decide and who participates? This speeds up processes enormously if you manage your meeting and decision-making platforms correctly.

Hornegger: What comes next?

Zipse: The third question is how do you communicate? The more you change, the more you have to communicate. People will follow as long as they know there is change. People will not follow if they do not know. This is all about internal communication, as well as external communication. We are very fortunate here because we have huge communication opportunities. Social media, regular press, internal media … and our team members, our employees, are multipliers of information. If you look at these three steps and take care that you do not leave any out, you will almost automatically be successful.

Hornegger: Is there still an ingredient missing?

Zipse: At the end of the day, of course, you have to love working with people. That applies equally to managers and university presidents. I see that here—

you know, you work with bright young minds, and you want to involve them. I think those are very important ingredients.

Now, to come back to your question. I see a lot of opportunities to work together, especially in the engineering section, where everything is a technical application. Industry is a technical application. In fact, there is a strong connection as we combine knowledge discovery and academic processes with engineering applications in industry. I think there is also a strong link in training, in education, and in industrial projects. We should strengthen this bond in general—not only in Germany. Of course, we also have links with universities in Asia and universities in the United States. But we can do that here in Bavaria, first of all. We have enormous knowledge at the FAU, and we should use it.

Hornegger: You mentioned communication. Digitalisation got a huge boost with the Covid-19 pandemic. I have noticed here at our university that the distances between people have been reduced through the use of digital technologies.

Zipse: How has this boost in digital technology made a difference?

Hornegger: We are in touch with our deans more often. I can contact students immediately if I see a post on social media, where I feel I should talk to them and understand what is going on. Is this something that you also experience within the BMW Group? And are you also in the same situation as we are: Even after the Covid-19 pandemic, we want to maintain some of these tools on a regular basis and use them to improve the overall communication with our people at FAU and our partners worldwide.

Zipse: Absolutely. It was really a watershed moment. Now you have the technology and the software is there to communicate even if you are far apart. What we found was: Distancing leads to proximity. The further away you are, the greater the urge to talk, to communicate. We will use that because communication strengthens the organisation—whether you are sitting in the same room or in separate rooms. Everyone feels that, with digitalisation, you can organise yourself better.

Hornegger: For example?

Zipse: Let's consider a workshop with 30 people. What do you normally do? You make a big introductory statement, and then you break into subgroups. Then everyone has to leave the room … and then you bring them back together and you do it again. If you use a modern videoconferencing system, it takes just one click to randomly divide the whole 30 people into six subgroups. You also know that you have exactly 15 minutes to discuss, and then you get them back. This is a hugely efficient tool for organising bigger groups. Didn't we know it before? Well, we knew it, but we were not quick to use it.

Hornegger: Sometimes you need a disruption to change.

Zipse: That is true. And this change offers a real potential, new ways of working together. On the other hand, we must not forget: True innovation requires personal interaction. I am very much convinced of that. So, overall, the future will be a mix of remote collaboration, working from home, mobile offices, and face-to-face meetings. Of course, the world will be different in terms of how we work together after the pandemic. I am quite sure of that, and I see it as a positive step forward.

Hornegger: Speaking of moving forward, let us come back to mobility. This is one area in which we are very strong in Bavaria. We see it at the universities when we look at the education programmes and at the research activities, as well as in the automotive industry in Bavaria. When you look to the future: What is your vision for sustainability and mobility in the future?

Zipse: Individual mobility is a private industry worldwide. It's different from public transport, which is not organised privately but by governments or municipalities. So linking individual mobility choices with societal demands raises issues of sustainability, especially when we look at the use of space. There is nothing sustainable about traffic jams, you know? I think the car industry has an important role to play in finding a solution to these issues, irrespective of the fact that it is privately organised and organised by market principles. And I think the companies of the future will be able to combine these two requirements: to have a profitable business model, to be privately organised according to market mechanisms and at the same time to contribute to the needs of society.

Hornegger: So, our current approach to mobility needs to change?

Zipse: We are not defenders of the current state of mobility. Nobody at BMW is a supporter of traffic jams. We want to have a solution where no one is stuck in traffic jams. So, I think that we have to bridge the gap between private mobility, being privately organised, having a profitable business model and being a member of society who provides solutions for society. Understanding and addressing this societal part becomes extremely important because almost everyone feels the negative effects of individual mobility. So, it's our job to provide solutions to this obvious conflict that we face.

Hornegger: Well, you've drawn the arc from basic development in your company to societal issues, and that's a perfect fit for FAU. We are a full-spectrum university that covers this wide range of different areas that you have mentioned. Oliver Zipse, thank you very much for this inspiring conversation.

Zipse: Thank you very much.

1.4 Outline of the Book

For the authors of the chapters in this book, the Road to Net Zero is a shared journey that requires discourse and learning. Guided by the same North Star, there are different, sometimes even competing, views on the precise route. Given the ambition of the goal and the complexity of the terrain, no single actor has the perfect solution. Therefore, the journey together requires joint efforts, the search for a balance between different objectives and smart ideas rather than pre-determined answers. It was with this in mind that the expert discussions between FAU and BMW, which form the core of each chapter in the book, were conducted. In designing the book, we were aware that we could not claim to be able to cover all the topics related to the sustainable transformation of organisations and companies towards Net Zero. Rather, the selection of topics reflects the thought processes that emerged from the expert conversations and that we would like to share with you.

Furthermore, the authors believe that the Road to Net Zero should not be understood as a linear process but rather as an iterative one, where each imperfect step, each iteration, represents an improvement in climate change mitigation. Similar to innovation processes, the overall vision is approached step by step, and the achievement of each iteration marks the beginning of a new cycle. Again, this is reflected in the choice of topics and the structure of the book, as shown in Fig. 1.1.

Overall, the Road to Net Zero, as described in this book, is organised into three thematic clusters. Chapters 2–4 deal with issues that mainly concern corporate strategy, from setting sustainability goals to integrated strategy formulation and integrated reporting. Chapters 5–7 deal with the operational aspects of an OEM (Original Equipment Manufacturer or carmaker) in the automotive sector, where, in addition to product development, the upstream and downstream supply chains play an increasingly important role, as does carbon-neutral production. The third and final group of topics, in Chap. 8, looks at the technological developments that will significantly shape and drive the transformation of the automotive industry in the future. The final chapter concludes the book with a management summary and a research agenda.

After decades of climate monitoring and climate impact research, and with the growing awareness of the immense challenge facing humanity, the Paris Agreement represents the most important agreement to date showing how all nations can work together to responsibly mitigate climate change in the future. The Paris Agreement was therefore chosen as the starting point for this book. Looking at the Paris Agreement from the perspective of an organisation

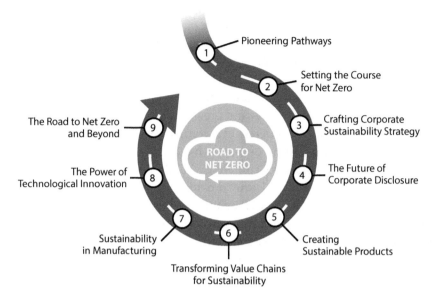

Fig. 1.1 Outline of the Road to Net Zero

or business, several questions arise: What are key climate science foundations that inform the Paris Agreement? How can the 1.5 °C target be achieved not only by sovereign states but also by individual companies and organisations? How can national climate targets be broken down to lower organisational levels to set sustainability targets that reflect the current scientific projections? How do you set targets as a company or organisation that wants to transform its business model and operations in a way that is credible and responsible to both the climate and its employees? These and other challenges to setting sustainability targets are reflected in the second chapter of the book, 'Setting the Course for Net Zero', by Markus Beckmann, Gregor Zöttl, Veronika Grimm, Thomas Becker, Markus Schober, and Oliver Zipse.

In the past, it was considered sufficient for any company to set sustainability goals separate from its corporate strategy—often referred to as corporate social responsibility measures. However, today's rapidly accelerating climate change requires a paradigm shift. Strategies that demonstrate a high level of maturity do not treat sustainability as a stand-alone add-on; rather, they integrate it into the way a company creates value. Achieving real lifecycle improvements requires integrated thinking that considers the entire value chain, not just the company's operations. This type of integrated approach to sustainability permeates the entire strategy process. The process extends from strategy formulation, which requires reliable target setting, through strategy implementation, which needs an integrated management approach, to strategy evaluation, which calls for new ways of measuring and reporting. In the third chapter, 'Crafting Corporate Sustainability Strategy', Markus Beckmann, Thomas Becker, and Oliver Zipse outline how integrated thinking changes the entire strategy process.

The statement 'What gets measured, gets done' reflects the fact that the Road to Net Zero for companies and organisations is decisively influenced by new ways of reporting. As interest in a company's sustainability strategy and performance grows, reporting on purely financial indicators is no longer sufficient to satisfy all stakeholder interests. While traditional reporting is primarily aimed at investors and thus provides information on the company's financial performance, today's companies require a broader focus on non-financial, sustainability-related aspects to meet the information needs of other stakeholders, such as employees, governments, or society. The transition to non-financial (sustainability) reporting has gradually evolved from voluntary standards with poor comparability to regulatory requirements for greater transparency. In the fourth chapter, 'The Future of Corporate Disclosure', Thomas M. Fischer, together with Oliver Zipse, Jennifer Adolph, Jonathan Townend, and Markus Schober, examine the transition from conventional to integrated reporting, while reflecting on recent legislation, the challenges of measuring and selecting non-financial and financial key performance indicators (KPIs) and the balancing of different stakeholder interests.

Reliable and credible sustainability targets, an integrated strategy and integrated reporting provide the roadmap for the path to Net Zero. With the introduction of electric vehicles, the majority of lifecycle emissions will shift from the use phase to the production phase. As a result, circular value chains play a key role in the operational transformation towards Net Zero and thus determine the second cluster of topics in the book. The transition to a circular economy starts with a new way of thinking about product development. Together with Oliver Zipse and Lena Ries, Sandro Wartzack opens the book's second thematic cluster with Chap. 5, 'Creating Sustainable Products'. The authors reflect on design for recycling, the replacement of scarce resources with secondary materials, and the introduction of natural materials, especially in interior design. Changes in consumer behaviour and the appearance of products with eco-efficient footprints are also discussed.

The discussion on sustainability in product development leads to the challenge of sourcing scarce and valuable resources. In particular, battery manufacturing and electric drivetrain manufacturing require materials that are available only in limited quantities and from only a few countries around the world. This increases the need for closed-loop supply chains where secondary materials can enter production. In addition, the Road to Net Zero depends heavily on suppliers far upstream in the supply chain, as science-based targets create responsibilities for the OEM throughout the entire supply chain. In addition to eco-efficient sourcing, the social dimension of raw material production must also be considered. In the sixth chapter, 'Transforming Value Chains for Sustainability', Kai-Ingo Voigt, Lothar Czaja, and Oliver Zipse reflect on these diverse challenges and show ways forward with practical examples from BMW and suggestions for future research.

Following the value chain downstream, a green factory can be seen as another crucial step in the transformation towards Net Zero at the operational level. Optimising production has been the focus of researchers and practitioners for more than two decades. Today, there is a consensus that sustainable manufacturing must cover the three dimensions of economic, environmental, and social aspects. While in the past the shift towards operational excellence was mainly driven within a factory, an integrated strategy approach now requires consideration across system boundaries again. The BMW iFactory is an example of a value-added network that is not simply a new production facility but combines lean systems, digitalisation technologies, and circular production processes to address the three dimensions of sustainable production. In the seventh chapter, 'Sustainability in Manufacturing', Nico Hanenkamp and Oliver Zipse together discuss the fundamentals of sustainable production and the latest advances in energy supply, circular processes, and manufacturing technologies.

Certainly, technological innovation provides new opportunities and impetus for further transformations on the Road to Net Zero. For this reason, Chap. 8, 'The Power of Technological Innovation', forms the third thematic cluster of this book and the ending/starting point of the Road to Net Zero, illustrated in Fig. 1.1. Technological innovations can trigger new strategies and goal adjustments that can take the continuous improvement cycle into a new round. The authors Jörg Franke, Peter Wasserscheid, Thorsten Ihne, Peter Lamp, Jürgen Guldner, and Oliver Zipse systematically analyse the drivetrains of the future. From the electric drivetrains to synfuel internal combustion engines (ICEs) and fuel cells, the authors discuss challenges and opportunities for each technology and outline possible future developments.

Ultimately, just as the development of new technological innovations depends on collaboration, so does an organisation's overall effort to move towards Net Zero. Therefore, in addition to a research agenda, the final chapter of the book, 'The Road to Net Zero and Beyond,' authored by Markus Beckmann and Irene Feige, initially provides a summary of the preceding chapters, discussing shared themes and insights. It then extends its discussion beyond the Road to Net Zero, and in the outlook, delves into the relevance and value of university–industry partnerships, accentuating the importance of collaborative efforts in achieving Net Zero objectives. It explores innovative forms of collaboration to address this complex and time-critical global challenge and to jointly identify strategic pathways for sustainability-driven business transformation in the automotive industry.

Are you ready to join us on the Road to Net Zero? We would love to take you on our journey so that we can grow and learn from each other.

Open Access This chapter is licensed under the terms of the Creative Commons Attribution 4.0 International License (http://creativecommons.org/licenses/by/4.0/), which permits use, sharing, adaptation, distribution and reproduction in any medium or format, as long as you give appropriate credit to the original author(s) and the source, provide a link to the Creative Commons license and indicate if changes were made.

The images or other third party material in this chapter are included in the chapter's Creative Commons license, unless indicated otherwise in a credit line to the material. If material is not included in the chapter's Creative Commons license and your intended use is not permitted by statutory regulation or exceeds the permitted use, you will need to obtain permission directly from the copyright holder.

2

Setting the Course for Net Zero

Translating Climate Science into Political and Corporate Targets

Markus Beckmann, Gregor Zöttl, Veronika Grimm, Thomas Becker, Markus Schober, and Oliver Zipse

2.1 Introduction

In 2015, a historic accord was reached in Paris, uniting 195 nations and the European Union in a collective commitment to climate action. The Paris Agreement set an ambitious target—limiting global warming to well below 2 °C compared to pre-industrial levels (Paris Agreement, see United Nations (UN), 2015). Despite the ensuing struggles with implementation and the limitations inherent in such a broad international pact, the Paris Agreement represents a monumental global breakthrough, embodying the goal of attaining net zero emissions and providing a roadmap for joint international environmental stewardship.

This book serves as an exploration of that roadmap and how businesses can contribute to it by charting the path known as the 'Road to Net Zero'. Therefore, as the first thematic focus of this book, Chap. 2 sets out the broad

---

Veronika Grimm contributed to this chapter exclusively through the expert conversation in Sect. 2.4.

M. Beckmann • G. Zöttl (✉) • V. Grimm
FAU Erlangen-Nürnberg, Nuremberg, Germany
e-mail: gregor.zoettl@fau.de

T. Becker • O. Zipse
BMW AG, Munich, Germany

M. Schober
FAU Erlangen-Nürnberg, Erlangen, Germany

context and objectives of this journey, outlining the background for sustainability transitions steered by businesses towards a climate-friendly future. While subsequent chapters examine the role of businesses in detail, this chapter first focuses on the interplay of climate science, policymakers, and corporations needed in setting the course for a decarbonised future.

The structure of the chapter is designed to guide readers through the multifaceted aspects of this complex topic. Section 2.2 lays the foundation with a brief overview of fundamental climate science. A basic understanding of these fundamentals is crucial to understanding the urgency and scope of the task at hand. Section 2.3 charts the evolution of global climate policy, taking readers along the path that led to the Paris Agreement. It elucidates why the pursuit of net zero emissions necessitates a profound shift away from current mainly fossil-fuel-based economies towards a sustainable, low-carbon future. Section 2.4 zooms in on the role of national and supranational policymakers. Their role in crafting regulations and incentivising changes is vital in propelling the world along the Road to Net Zero. Engaging with these basics, Sect. 2.5 features the expert conversation of Prof. Dr Veronika Grimm, FAU Chair of Economic Theory and Member of the German Council of Economic Experts, Prof. Oliver Zipse, CEO of BMW Group, and Dr Thomas Becker, VP Sustainability & Mobility at BMW. They shed light on the intricate balancing act between regulation, infrastructure support, and the strategic and technological imperatives of businesses. Finally, Sect. 2.6 delves into the science-based frameworks for setting, measuring, and reporting climate targets at the corporate level in line with the Paris Agreement before Sect. 2.7 concludes.

2.2 A Brief Review of Selected Climate Science Insights

A basic understanding of climate science fundamentals is helpful for grasping the complexities of climate change and the pressing need for mitigative action. The aim of this section is to encapsulate some of these rudiments, especially those relevant to the Road to Net Zero, as elaborated upon in this book. While the idea of 'following the science' is indeed crucial, it is important to acknowledge that science is an iterative field that progresses through a constant exchange of ideas and testing of theories. Science does not provide static answers; rather, it generates ever-evolving explanations that are refined over time based on new evidence and understanding. This concept of iterative refinement and learning applies equally to climate science.

The Intergovernmental Panel on Climate Change (IPCC) exemplifies this process of scientific collaboration and consensus building. Composed of three working groups, the IPCC does not engage in original research; instead, it synthesises the global body of climate research to provide policymakers and the general public with comprehensive assessments of the scientific consensus on climate change (IPCC, 2021b). Working Group I assesses the physical science basis of climate change; Working Group II addresses the vulnerability of socio-economic and natural systems to climate change, the negative and positive impacts of climate change, and options for adapting to it; and Working Group III assesses options for reducing greenhouse gas emissions and otherwise mitigating climate change (IPCC, 2021a).

Central to understanding climate change is recognising the role of carbon dioxide (CO_2) and other greenhouse gases (GHGs) in heating our planet. CO_2 and its equivalents trap heat in the Earth's atmosphere, thus affecting global temperatures (Shakun et al., 2012). Prior to the first industrial revolution, the emissions and absorptions of greenhouse gases were in balance, resulting in relatively limited alternating CO_2 concentrations and temperatures. However, increasing anthropogenic (i.e. human-made) actions, such as industrial and economic activities, have disrupted this delicate balance of the global carbon cycle. The burning of fossil fuels releases carbon from the ground into the atmosphere, while land use changes (such as deforestation and the conversion of wetlands into agricultural land) also emit carbon and reduce the Earth's natural capacity to absorb carbon. As a result, the concentrations of CO_2 in our atmosphere have been rising and, in turn, have been warming the planet (IPCC, 2023; Prentice et al., 2012).

Accurately measuring the historical and current levels of GHG emissions, including CO_2, is a complex but crucial part of climate science. By 2022, atmospheric CO_2 concentrations had reached around 421 parts per million (ppm), more than 50% higher than pre-industrial levels (National Oceanic and Atmospheric Administration, 2022). This increase in CO_2 concentrations corresponds to a warmer planet, with the average surface temperature estimates from 2011 to 2020 indicating the Earth was already approximately 1.1 °C warmer than during the pre-industrial period (1880–1900) (IPCC, 2023, p. 4).

Climate science has harnessed such historical data to construct and validate advanced simulation models for predicting future temperature levels. In IPCC reports, these future simulations rely on varying emission scenarios, also known as shared socio-economic pathways. These scenarios contemplate different potential socio-economic and technological developments. For each scenario, simulation models can anticipate future GHG emissions and the corresponding shifts in global temperature (IPCC, 2021c).

Comprehending the spectrum of these temperature variations is vital for assessing the potential ramifications of climate change. These consequences are extensive, permeating almost all aspects of our lives. They encompass increased frequency and intensity of heatwaves, extended droughts, unpredictable precipitation, escalating biodiversity loss, intensified wildfires, invasive species, forest loss, rising sea levels, melting ice caps and glaciers, ocean acidification, vanishing coral reefs, heat-related illnesses and mortality, the spread of vector-borne diseases, and heightened food insecurity (IPCC, 2022).

Given these threats, a protracted discourse has emerged regarding the critical levels of global warming beyond which humanity would face severe and dangerous climate change. Within the IPCC, two such goals have gained specific attention: the 2 °C threshold and the more recent 1.5 °C target. The 2 °C target, first proposed in the 1970s by economist William Nordhaus, later garnered political recognition in the 1990s. By the time of the IPCC's Fourth Assessment Report (AR4) in 2007, this target had become a commonly referenced goal in policy discussions. However, the AR4 did not explicitly endorse the 2 °C target but instead presented a range of possible outcomes based on different emission trajectories. It highlighted that a global temperature rise of 2 °C above pre-industrial levels would have serious impacts, including an increased risk of extreme weather events, significant biodiversity loss, and a higher likelihood of tipping points in the Earth system (IPCC, 2007). The IPCC's Fifth Assessment Report (AR5), published in 2014, further reinforced these findings (IPCC, 2014). Consequently, the IPCC was tasked with preparing a special report on the impacts of global warming of 1.5 °C. This report, published in 2018, clarified that the impacts at 1.5 °C of warming are significantly less than at 2 °C and underscored the need for rapid, far-reaching, and unprecedented changes in all aspects of society to achieve this more ambitious target (IPCC, 2018).

In its most recent synthesis report, published in 2023, the IPCC underscored the significance of this shift towards the 1.5 °C goal because the latest stage of climate science suggests that dangerous forms of global warming are likely to occur at lower levels of global warming than previously anticipated 'due to recent evidence of observed impacts, improved process understanding, and new knowledge on exposure and vulnerability of human and natural systems, including limits to adaptation' (IPCC, 2023, p. 15). One specific concern is the potential for the climate system to reach tipping points and trigger self-enforcing feedback loops. Such tipping points, including the melting of Arctic sea ice or the thawing of permafrost (which releases methane), could contribute to further warming, even if anthropogenic emissions were fully eliminated.

The quest to prevent these catastrophic impacts led the IPCC to study the probable effects of limiting global warming to 2 and 1.5 °C above pre-industrial levels. Their analyses provide the scientific basis for these temperature goals, which are now central to global climate policy.

Emerging from this research is the concept of a 'carbon budget'. This is the aggregate amount of CO_2 emissions that can be discharged into the atmosphere while still maintaining a likely chance of limiting global warming to a specific temperature target. The size of the remaining carbon budget varies depending on whether the goal is to limit warming to 2 or 1.5 °C. In its latest synthesis report, the IPCC estimates the remaining carbon budgets from the beginning of 2020 to be 500 Gt CO_2 (for a 50% likelihood of limiting global warming to 1.5 °C) and 1150 Gt CO_2 (for a 67% likelihood of limiting warming to 2 °C) (IPCC, 2023, p. 21). At the 2019 emissions level, this budget would be almost fully utilised for the 1.5 °C goal and roughly a third would be utilised for the 2 °C goal by 2030.

Therefore, climate science illustrates that achieving both the 2 °C and the 1.5 °C goals is feasible only with a massive and rapid decarbonisation of the economy. Regardless of whether the target is to limit warming to 2 or 1.5 °C, global emissions must reach 'net zero' at the culmination of this process. This term implies that any emissions discharged into the atmosphere must be offset by equivalent removals, either through natural processes (i.e. by absorption by natural sinks, such as plants and the ocean) or human-made technologies, such as carbon capture and storage.

However, a crucial point to emphasise is that achieving net zero emissions by a specific year is not sufficient to stay within a specified carbon budget. What matters for climate stabilisation is the accumulated emissions over time, which means the actual reduction pathways on the way to net zero. This means that if emissions are reduced too slowly in the early years, then faster reductions will be needed later to stay within the carbon budget. Thus, while a net zero target sets the end goal, the pace at which emissions decrease on the Road to Net Zero is just as crucial (IPCC, 2021c).

In summary, understanding the basics of climate science is key to appreciating the challenges of climate change and the urgency of taking action to mitigate its worst impacts. As science evolves, so too must our responses to it. The Road to Net Zero, for example, is not just about reaching a destination; it is also about how swiftly we embark on that journey and how many iterations (cf. Chap. 1; Fig. 1.1) we will need to do so. The following section discusses the evolution of this journey in the global climate policy debate.

2.3 Global Climate Policy: The Road to Paris and Beyond

The origins of global climate policy can be traced back to the 1972 United Nations Conference on the Human Environment and the ensuing establishment of the United Nations Environment Programme (UNEP). The conference, hosted in Stockholm, provided the foundation for international environmental cooperation (Bodansky, 2001). Recognising the potential threat of climate change, UNEP, in collaboration with the World Meteorological Organization (WMO), formed the Intergovernmental Panel on Climate Change (IPCC) in 1988. This independent entity was tasked with assessing scientific literature and furnishing crucial scientific information to the climate change process.

The Earth Summit of 1992 in Rio de Janeiro heralded the establishment of the United Nations Framework Convention on Climate Change (UNFCCC)—a pivotal international treaty devoted to addressing climate change. The UNFCCC, grounded in scientific insights suggesting that human-made greenhouse gas (GHG) emissions could influence global temperatures, was adopted with the ultimate objective of preventing 'dangerous anthropogenic interference with the climate system' (UNFCCC, see UN, 1992, Article 2). While the UNFCCC aimed to stabilise atmospheric GHG concentrations to preclude dangerous warming, it did not specify the level at which this stabilisation should occur.

The UNFCCC came into effect on March 21, 1994, and today enjoys near-universal membership. The 198 nations that have ratified the Convention regularly convene for global climate conferences, referred to as the Conferences of the Parties (COPs). These ongoing conferences highlight how the history of global climate policy has been shaped by the dynamic interplay between emerging insights from climate science and political negotiations on a global scale. Climate policy is informed not only by findings from climate science but also by political evaluations and decisions that extend beyond the realm of science (e.g. when discussing what impacts count as dangerous or how burdens should be distributed).

Political negotiations concerning climate change have encompassed a broad array of topics. As the impacts of climate change become increasingly evident, more attention is devoted to questions of adaptation, protection of the most vulnerable, assisting developing countries with the transition and debates about financial compensations for countries most affected by global warming. While these topics are pertinent and relevant for international political negotiations, the following review is specifically focused on the policy discussion on limiting global warming, the emergence of global science-informed targets and, consequently, on selected milestones for setting the trajectory for the Road to Net Zero discussed in this book.

The Kyoto Protocol, adopted in 1997 at the third Conference of the Parties (COP 3) in Kyoto, Japan, marks the first significant milestone in global climate policy. For the first time, it introduced legally binding obligations for developed countries to reduce GHG emissions, thereby sparking international cooperation on climate change mitigation. These so-called Annex I countries (including the EU, Canada, Australia, New Zealand, and Russia) committed to substantial reductions in their greenhouse gas emissions—a 5% reduction compared to the 1990 level, with the target period set between 2008 and 2012. However, countries like China, India, Brazil, and Indonesia ratified the treaty without agreeing to binding targets. By establishing the principle that nations bear common but differentiated responsibilities concerning climate change, the Kyoto Protocol laid essential groundwork for subsequent climate agreements, including the Paris Agreement (Bodansky, 2001).

From a more technical perspective, the Kyoto Protocol is also noteworthy for its definition of the most relevant greenhouse gases (the seven Kyoto gases), encompassing not just CO_2 but also methane (CH_4), nitrous oxide (N_2O), hydrofluorocarbons (HFCs), perfluorocarbons (PFCs), sulphur hexafluoride (SF_6), and nitrogen trifluoride (NF_3). These latter six gases have their warming potential translated into CO_2 equivalents when determining emissions and emissions reductions. The Kyoto Protocol is also notable for its adoption of market mechanisms, such as emissions trading, the clean development mechanism (CDM), and joint implementation (JI). These innovative tools provided flexibility in how countries could fulfil their commitments (Barrett, 2005). Despite this, the Kyoto Protocol faced criticism for various limitations, including the absence of some major emitters (including the USA, which refused to ratify) and challenges in achieving its targets (Barrett, 2005).

While the Kyoto Protocol defined reduction targets for GHG emissions, it did not include a specific temperature or GHG concentration target to specify what 'dangerous interference' with the climate system implies. This began to change with the Copenhagen Accord, which was developed during the 15th Conference of the Parties (COP15) in Copenhagen in 2009. The Copenhagen Accord, reflecting advances in climate science and political assessments of climate impact, was the first instance of a global temperature target being explicitly mentioned in an international climate policy document. The Accord stated that 'deep cuts in global emissions are required […] to hold the increase in global temperature below 2 degrees Celsius' (Copenhagen Accord, see UNFCCC, 2009, p. 5). Nevertheless, a notable point is that the Copenhagen Accord did not formally establish the 2 °C target and lacked legal bindingness. The 2 °C target was officially adopted the following year, at COP16 in Cancun, which culminated in the Cancun Agreement.

Arguably, the most significant milestone in charting the Road to Net Zero thus far is the Paris Agreement, enacted at the subsequent COP17 in 2015. The Paris Agreement not only reaffirmed the 2 °C goal but pushed further, aiming to limit global warming to 1.5 °C if possible. While the Paris Agreement encompassed various important aspects, such as matters of adaptation, loss and damage, and climate finance—thereby emphasising the need for responses to climate change to extend beyond mitigation efforts alone (Klinsky et al., 2017)—the following features and implications of the Paris Agreement are particularly relevant for the Road to Net Zero, as they specify its purpose, destination, group of travellers, and travel model.

First, in terms of the purpose of the Road to Net Zero, the Paris Agreement (see UN, 2015, p. 3) established a global commitment to prevent dangerous climate change by keeping global warming 'well below 2 °C above pre-industrial levels and pursuing efforts to limit the temperature increase to 1.5 °C above pre-industrial levels'. The Paris Agreement thus represents a significant, though still insufficient, step forward in recognising the urgency of the climate crisis (Rogelj et al., 2016), with the inclusion of the 1.5 °C goal reflecting current climate science insights regarding the risks associated with warming above this threshold.

Second, in relation to the common destination of the journey to Net Zero, the Paris Agreement was the first global treaty to complement a temperature goal with a long-term goal of achieving net zero emissions by the latter half of the century. This goal of achieving net zero emissions had not been explicitly included in global climate agreements prior to the Paris Agreement. The Kyoto Protocol and other earlier agreements focused primarily on setting specific, near-term targets for reducing greenhouse gas emissions from developed countries, rather than stipulating a long-term global goal of achieving net zero emissions. However, while the concept of net zero emissions has become central to discussions on how to achieve the temperature goals of the Paris Agreement; notably, the phrase 'net zero emissions' does not appear verbatim in the text of the Agreement. Instead, the Agreement states that Parties aim to reach 'a balance between anthropogenic emissions by sources and removals by sinks of greenhouse gases' in the second half of the century (Paris Agreement, see UN, 2015, p. 4).

Third, regarding the group of travellers embarking on the journey to Net Zero, the Paris Agreement includes commitments from all countries to reduce their emissions. Unlike the Kyoto Protocol, which legally mandated emission reductions for developed countries only, the Paris Agreement brought together all countries, including major emitters like the United States and emerging economies such as China and India. The Agreement requires all nations, regardless of development status, to set, report on and revise their climate goals. This global responsibility reflects the rising emissions from developing countries and underscores a shared duty for climate action (Bodansky, 2016). It also explicitly states that the journey to Net Zero is a challenge of a global nature.

Fourth, in terms of the mode of travel, the Paris Agreement defines the journey to Net Zero as an iterative, continuous learning process. Contrasting with the Kyoto Protocol's top-down approach, where international targets were imposed and enforced on nations, the Paris Agreement encouraged a bottom-up approach. In this arrangement, each country develops its own climate plan—the so-called Nationally Determined Contributions (NDCs)—detailing its emission reduction targets and adaptation strategies. Intended to foster flexibility and national ownership of climate commitments, the concept of the NDCs is to engage countries in a regular review process, in which each country's NDCs are reviewed and updated every 5 years in what is known as the 'Global Stocktake' (Paris Agreement, see UN, 2015, pp. 18–19).

The regular review process of the Paris Framework is intended to ensure that efforts to address climate change are progressively scaled up. Despite this intention, critics argue that without a legally binding enforcement mechanism, the Global Stocktake relies excessively on international peer pressure and goodwill to drive increases in ambition, with a lack of legally binding mechanisms to enforce climate action (Bodansky, 2016). Irrespective of this criticism, an important point to note is that, despite its historic character, the Paris Agreement was not a breakthrough that resolved all issues. On the contrary, the Paris Agreement marked the beginning of continuous follow-up negotiations, with the operational details for the practical implementation of the Paris Agreement agreed upon at the UN Climate Change Conference (COP24) in Katowice, Poland, in December 2018—colloquially referred to as the Paris Rulebook—and finalised at COP26 in Glasgow, Scotland, in November 2021.

So, what are the key implications of the Paris Agreement for the Road to Net Zero as discussed in this book? It is difficult to overstate that Paris represents a fundamental departure from previous agreements by defining the long-term net zero-emission goal instead of short-term incremental reductions. The net zero goal is an absolute target that differs qualitatively from relative reductions. To achieve this distinct target, merely improving the efficiency of fossil-fuel-based technologies will not be enough. Instead, companies and the entire economy need a radical transition towards extensive decarbonisation. This necessitates a shift in energy sources, technologies, products, value chains, infrastructure, regulation, and much more.

Moreover, by affirming the global commitment to the 2° goal and the ambition to achieve the 1.5 °C goal, the Paris Agreement has anchored the concept of a remaining carbon budget in the global policy discourse. This implies that not only the long-term Net Zero goal matters but that ambitious reduction pathways are needed to align with Paris. This requires rapid and far-reaching emissions reductions.

The Paris Agreement's commitment to a bottom-up approach invites individual countries to formulate and submit their NDCs. This indicates that paths to Net Zero may differ between countries. So far, however, these national pledges are far from sufficient. In fact, a report by the UNFCCC in 2022 warned that the combined climate pledges of all 193 parties to the Paris Agreement would result in about 2.5 °C of warming by the end of the century (UNFCCC, 2022). More ambition is therefore needed. Consequently, with countries needing to step up their efforts, the spotlight now falls on the climate policies that can be enacted at the national level, a topic the subsequent Sect. 2.4 delves into. Meanwhile, multinational corporations aiming to align their operations with the Road to Net Zero must effectively navigate this complex policy terrain. Science-based frameworks can assist these companies in developing long-term strategies that consider diverse national climate policies. This is particularly pertinent, as the Paris Agreement demonstrates the evolutionary nature of global climate policy, reflecting the ever-evolving insights provided by climate science. Therefore, taking an active role on the Road to Net Zero and preparing for prospective regulations are greatly aided by a science-based approach, a topic that we will examine further in Sect. 2.6.

2.4 The Role of National Policy Frameworks and Governance Mechanisms

As discussed above, the Paris Agreement employs a bottom-up approach, encouraging all nations to propose their NDCs. Internationally, this necessitates negotiations on how these national efforts collectively align with the vital emission reductions needed to adhere to the 2 or 1.5° goal (Rogelj et al.,

2016). Conversely, on a national level (and supranational level, in the case of the EU), the challenge extends beyond setting ambitious emission reduction targets to formulating the domestic policies that will facilitate these reductions.

Accordingly, national governments and supranational entities like the European Union find themselves navigating a complex and challenging landscape. On the one hand, they must determine their fair share of the globally agreed climate goals, even in the absence of sanctions for non-compliance. They must also devise policy instruments that can effectively catalyse rapid and significant emission reductions. On the other hand, they must simultaneously consider the potential costs associated with these mitigation measures, their impacts on local populations and their implications for economic innovation and competitiveness.

In this situation, policy frameworks and governance mechanisms that promote sustainable innovations and offer significant economic value for domestic companies are highly desirable, even in the absence of clear implementation roadmaps and sanction mechanisms. Conversely, it becomes much more challenging to justify other climate measures that, while crucial to upholding a fair national contribution to the agreed-upon climate goals, impose substantial mitigation costs at the national level and do not provide dynamic benefits in terms of innovativeness and competitiveness for domestic companies. Therefore, domestic policymakers strive to enact climate policies that balance ecological effectiveness, economic efficiency, and legal and administrative feasibility while maintaining political acceptability.

The subsequent analysis will spotlight the actions taken by the EU and Germany as examples of concrete policy measures employed to balance these diverse objectives.

At the core of the EU's climate policy plans lies the European Green Deal, which was proposed and introduced in 2019 (see European Commission, 2023a). It includes the goal of reducing net greenhouse gas emissions by at least 55% compared to 1990 levels by 2030 and provides a roadmap for transforming the EU into a carbon-neutral continent by 2050. The Green Deal encompasses a wide range of specific goals, such as the sustainable use of resources (known as the Circular Economy Action Plan; see European Commission, 2023c), as well as specific sustainability goals and emission targets for different sectors, including the mobility and the building sectors (see European Commission, 2023d). However, in addition to these emission reduction goals, the Green Deal explicitly aims to promote innovation and competitiveness among domestic firms and industries by fostering the development of products and markets for clean technologies. These goals have been enshrined in the European Climate Law, which came into force in July 2021

(European Commission, 2023f; European Climate Law, see European Parliament & European Council, 2021). Nonetheless, the ultimate responsibility for implementing most of these ambitious policy goals lies with the individual national governments within the EU.

Governments have a wide range of policy tools at their disposal to implement specific emission reduction targets. Here, we provide an overview of the most important and prominent instruments adopted in Germany. We categorise these instruments into three groups: (1) market-based instruments that assign a proper price to external costs; (2) direct support instruments designed to promote the development and adoption of sustainable technologies and products, including infrastructures; and (3) traditional regulatory approaches, also known as 'command-and-control measures', which involve the direct prohibition of specific polluting technologies or products.

2.4.1 Market-Based Instruments to Directly Internalise External Costs

The fundamental idea behind these policy tools is to impose a market price on activities that have a detrimental impact and cause damages borne by society as a whole, rather than directly affecting producers or consumers involved. External costs occur when the cost of these damages is not fully reflected in the market prices faced by the market participants directly involved. Market-based instruments are designed to correct this mismatch and impose prices that properly reflect the external costs caused (Pigou, 1920).

Two popular market-based mechanisms are available for reducing GHG emissions: carbon taxes and cap-and-trade systems (Baumol & Oates, 1988). A carbon tax imposes a fee per unit of emissions, encouraging businesses to reduce emissions to lower their tax burden. This mechanism sets a certain price on emissions, but the emissions reduction is uncertain. Conversely, cap-and-trade mechanisms set a firm limit on total emissions (the cap). Entities can then buy and sell emissions allowances (the trade), which provide certainty on emissions reduction but variability in cost. While a carbon tax eliminates carbon price volatility, cap-and-trade ensures meeting the target, but at the disadvantage of volatile carbon prices (Tietenberg, 2006).

An important policy instrument of this kind is the European Union Emissions Trading System (EU ETS). The system was launched in 2005 as a cap-and-trade mechanism with the aim of pricing GHG emissions and limiting total emissions (for detailed information on the functioning of the EU ETS, see, for example, European Commission, 2023e). Currently, the EU

ETS covers emissions from electricity production, energy-intensive industries (such as iron, steel, cement, glass, etc.), and some parts of aviation and maritime transport. Overall, the EU ETS covers approximately 40% of the current GHG emissions in the EU (European Commission, 2023e). In December 2022, the EU Parliament and the EU Council agreed to strengthen the EU ETS.

Also in December 2022, the EU Parliament and the European Council agreed to extend the ETS to emissions occurring in the transport and building sectors that were so far not included in the EU ETS (European Parliament, 2022). Germany had already introduced a national CO_2 price for these sectors. The corresponding law (Brennstoffemissionshandelsgesetz, see Deutscher Bundestag, 2022a) was introduced in 2019 and became effective in 2021. This system is phased in by a period of yearly increasing emission prices and then transitions into a cap-and-trade system. The prices for CO_2 emissions increase in several steps from 25 €/ton CO_2 in 2021 to 45 €/ton by 2025, which correspond to 0.07 and 0.11 €/l gasoline, respectively (see, for example, Umweltbundesamt (UBA), 2022). In this early phase, the system thus implements de facto a CO_2 tax. From 2026 on, a switch to an emission trading system is planned, however within a price corridor between 55 and 65 €/ton CO_2 (Brennstoffemissionshandelsgesetz, see Deutscher Bundestag, 2022a).

Market-based instruments that aim to internalise the external costs of emissions can be powerful tools to combat climate change (Stavins, 2003). When properly established, emission targets or carbon taxes can lead to an efficient achievement of climate goals within a closed economy (Baumol & Oates, 1988). However, the reality is that neither the EU nor its member states operate in isolation. Climate change is a global problem, but market-based mechanisms are typically implemented on a relatively small, national, or European scale within the framework of open economies. In response to this, the EU proposed the EU Carbon Border Adjustment Mechanism (CBAM), a policy that mandates importers of specific goods into the EU to pay for the carbon emissions embodied in those goods. The CBAM is designed to prevent carbon leakage, a phenomenon that occurs when companies relocate their production to nations with less stringent climate policies (European Commission, 2023b).

Part of the EU's 'Fit for 55' package, the CBAM will apply to a range of products, including cement, iron and steel, aluminium, fertilisers, and electricity. Advocates of the CBAM anticipate that it could help diminish GHG emissions by encouraging other countries to adopt more rigorous climate policies. The mechanism also aims to shield EU industries from unfair competition arising from nations with laxer climate regulations. However, the

CBAM has faced its share of criticism. Some critics argue that its implementation could be challenging and potentially spark trade disputes with other countries or—if the coverage is incomplete—may lead to the relocation of value chains outside the EU (Garnadt et al., 2020; Sachverständigenrat zur Begutachtung der gesamtwirtschaftlichen Entwicklung, 2020). There are also concerns that the CBAM could disproportionately impact developing countries that depend on the export of carbon-intensive goods.

The CBAM is thought to address the challenges market mechanisms face when carbon prices vary across countries and sectors. In an ideal setup, there should be a global uniform carbon price for all market participants (Nordhaus, 2019). Through the thorough implementation of such mechanisms, we could in principle address climate emissions cost-effectively. The idea is that if emission targets are reliably announced, they could provide strong long-term incentives to trigger the necessary investments and spur innovation in technologies and infrastructure, such as green energy generation and hydrogen or electric mobility charging stations (Aldy et al., 2010).

However, global carbon prices are currently absent. Moreover, there can be additional market or government failures, which could arise from political uncertainty, imperfect financial markets, administrative and transaction costs, limited appropriability of innovative activities, or network effects. Failures may occur, for example, when governments unexpectedly adjust carbon prices for political reasons, when green start-ups struggle to secure venture capital, when network effects affect infrastructure, or when innovation rents cannot be fully appropriated.

Politicians, tasked primarily with the welfare of their domestic populations, navigate this complex terrain. While market-based instruments represent a powerful option to tackle climate change (Stern, 2007), integrating such mechanisms with other policy tools may be advantageous from both a national and a European perspective. This approach would help address the limitations of a single-policy method and potentially enhance the effectiveness and efficiency of climate change policies (Goulder & Parry, 2008).

2.4.2 Direct Support Instruments for Sustainable Technologies and Products

A second category of policy tools directly incentivises the adoption and development of technologies and solutions crucial for mitigating greenhouse gas emissions. Economically, this approach is especially sensible for nascent technologies. Early-stage technologies frequently confront challenges, such as high costs, infrastructure deficiency, and market uncertainty, that could impede their development (Jaffe et al., 2002). Government subsidies can help overcome these barriers, driving 'directed technical change' towards greener technologies that can not only correct market failures related to environmental externalities but also stimulate innovation and economic growth (Acemoglu et al., 2012). However, note that such subsidies must be carefully designed to ensure they are cost-effective and do not lead to unintended consequences.

A prime example of such an instrument is the German Renewable Energy Sources Act (Erneuerbare-Energien-Gesetz or EEG; see Bundesministerium für Wirtschaft und Klimaschutz, 2023). First introduced in 2000, the EEG has seen several revisions to adapt to changing market conditions and technological advancements. Its primary objective is to support the expansion of renewable energy generation and reduce the country's dependency on fossil fuels. The EEG provides a stable and long-term framework for the development of renewable energy projects by guaranteeing minimum compensations for electricity generated from renewable sources. Those guaranteed compensations typically are granted for a period of 15–20 years, for most renewable projects they are determined in tender procedures. The EEG covers a wide range of renewable energy technologies, including wind power, solar power, biomass, hydropower, and geothermal energy. It establishes specific compensation guarantees for each technology, also considering factors such as the installation size, technology type, and regional resource potential.

Largely as a result of the Renewable Energy Sources Act (EEG), the proportion of gross electricity consumption in Germany derived from renewable energy sources has seen a significant increase in recent years. The share rose to 41% in 2021 and further escalated to 46% in 2022 (Umweltbundesamt (UBA), 2023b). However, it is critical to recognise that the reported electricity consumption of 549 TWh in 2022 (UBA, 2023c) constitutes merely a fraction of Germany's total energy consumption, which approximates around 2500 TWh (UBA, 2023a). Currently, the largest portion of energy consumption is non-electric energy, which is predominantly utilised in the mobility and heating sectors. These sectors are expected to undergo major electrification in the coming years. While the electrification process is expected to considerably boost energy efficiency, and the current version of the EEG-2023 proposes highly ambitious expansion paths, especially for wind and solar power (Erneuerbare-Energien-Gesetz, see Deutscher Bundestag, 2023, §4), it still presents a formidable challenge to fully cover the drastically increased electricity needs with German domestic renewable energy sources.

Germany also makes serious efforts to promote sustainable products that are considered to contribute significantly to low-emission scenarios, particularly in the mobility sector. Since 2016, the German government has implemented a financial incentive programme called the 'Umweltbonus' to promote the purchase of battery electric vehicles (BEVs) and fuel cell electric vehicles (FCEVs). As of 2023, the programme provides subsidies of up to 4500 € for smaller cars and 3000 € for medium-sized cars. From 2024 onwards, these subsidies will be reduced, with only smaller cars eligible for purchase subsidies (Presse- und Informationsamt der Bundesregierung, 2022). Additionally, all newly registered BEVs and FCEVs are exempted from the vehicle tax until 2030 (equivalent to approximately 100–200 € per year; see Bundesministerium der Finanzen, 2023). Finally, since 2017, substantial support programmes have been introduced to incentivise the installation of private and publicly accessible charging points for BEVs throughout Germany with a total volume of 1100 Mio. €. Public support programmes for installing charging points for FCEV are also in place, focused on commercial vehicles and requiring the usage of 100% green hydrogen, currently resulting in significantly smaller support volumes as observed in the case of electric charging points (Bundesministerium für Digitales und Verkehr, 2021).

2.4.3 Traditional Regulatory Approaches: 'Command-and-Control Measures'

Finally, a third category of governance mechanisms, known as command-and-control measures, imposes mandatory regulations and standards to regulate emissions and promote sustainable practices (Tietenberg & Lewis, 2018). From an economic perspective, these instruments are typically perceived as the least efficient. However, under certain specific circumstances, command-and-control tools can be relevant and may even provide a superior alternative to market-based instruments. Factors such as transaction costs, administrative costs, possibilities for strategic behaviour, or political costs can influence this preference (Newell & Stavins, 2003). Similarly, mandatory uniform standards in environmental regulation might be economically justified and offer 'efficiency without optimality', for example, when transaction costs make market-based mechanisms impractical (Baumol & Oates, 1988, p. 159).

A prominent example of command-and-control instruments are the regulations and laws to realise the German Coal Phase-Out. The goal is to eliminate Germany's dependence on coal for energy production and transition to cleaner and more sustainable sources of power. The corresponding law has been effective since 2020 (Kohleverstromungsbeendigungsgesetz; see Deutscher Bundestag, 2022b). One essential feature of this law is the prohibition of constructing new coal-fired power plants in Germany. Additionally, it includes a precise timeline for phasing out existing coal-fired power plants with the aim of shutting down all coal-fired power plants by 2038 at the latest. This includes both hard coal (anthracite and bituminous coal) and lignite (brown coal) power plants. Owners of existing coal plants are compensated for exiting the market.

Command-and-control mechanisms also play a significant role in the regulation of emissions within the German and European mobility sectors. At the EU level, emission standards have been set to reduce the environmental impact of vehicles and improve air quality. These standards, known as EURO-Norms (currently EURO 6), limit the amount of pollutants, such as nitrogen oxides, particulate matter, and carbon monoxide, emitted by vehicles. Further regulations explicitly limit the average fleet emissions of the greenhouse gas CO_2. Since 2020, this has been governed by Regulation (EU) 2019/631 (see European Parliament and European Council, 2019). The standards have been progressively tightened over the years, with each new iteration introducing stricter emission and measurement limits. Recently, as part of the EU's 'Fit for 55' package, concrete plans are in place to enforce zero emissions for new passenger cars and vans from 2035 onwards. However, to date, the fleet emission targets pertain solely to tailpipe emissions and do not account for emissions related to manufacturing the vehicles. As the transition towards electric mobility shifts emissions from the use phase to the production phase, there is increasing momentum to consider all CO_2 emissions throughout the entire life cycle of those vehicles (see European Parliament, 2023).

The aforementioned compilation of policy frameworks and governance mechanisms represents the most prominent initiatives driving Germany's sustainable energy and mobility transition as described in this chapter. These policies and mechanisms play a central role in shaping the country's renewable energy landscape and fostering the integration of sustainable transportation solutions. However, it is crucial to acknowledge that this overview does not provide a comprehensive survey of all relevant measures implemented in Germany. The measures supporting and guiding towards a successful energy and mobility transition indeed encompass a diverse range of additional strategies and initiatives that all contribute to the ongoing transformation. In sum, policy measures across the spectrum, from market-based mechanisms to supportive subsidies and command-and-control regulations, all have strengths and limitations in managing environmental issues in practice. The preferability of each approach depends on the specific context and requires careful economic and policy analysis to determine the most efficient and effective combination of measures.

2.5 Expert Conversation on Economic Climate Policy: Between Technological Openness and Regulation

What Are the Challenges and Opportunities of Sustainability as a Corporate Strategy?

Grimm: This is a great opportunity to talk about sustainability in companies and also climate goals in the world. What is the biggest challenge for you at BMW, when you look at the climate goals that have now been tightened even in Germany and the EU?

Zipse: Sustainability is at the core of our operation here at BMW. At the same time, political circumstances change, and climate change is on the run. Everyone knows that this is one of the major challenges in the world. Our customers' behaviours are also changing. Therefore, it was about time to put sustainability right at the centre of our corporate strategy. That is for customer reasons, that is for political reasons and also for economic reasons.

Grimm: For decades, companies have regarded sustainability as a necessary and costly exercise. What has led to this shift in focus?

Zipse: If we neglected sustainability due to commercial reasons, our strategy would not be sustainable. Think about the increasing prices of resources. If we do not include the possibility of reusing materials through recycling into our corporate strategy, we could encounter difficulties producing cars as we do today in the near future. That is why we have created a new architecture for new cars, which is going to be launched in 2025 and puts the secondary use of materials at the core of the architecture. What we see today is that steel, copper, nickel, palladium, and rhodium are becoming increasingly more expensive. Hence, we must find ways to keep cars affordable for our customers.

Do Politics and Industry Approach Sustainability from Different Perspectives?

Grimm: In a way, one can get the impression that the industry is now taking the side of the people who protect the climate and have fought for a long time—also in the streets—against climate change much more ambitiously than the politicians are doing at the moment. At least, it seems that industry can be faster in some way. Politicians have to fight for goals, have to fight for the implementation of certain regulations. Are the regulatory frameworks already working? Do companies have the right incentives to advance towards sustainable business?

Zipse: I think it is good advice to have very close contact and a trusted relationship to politicians, because we actually share the same framework about our future path. Politicians are elected by members of society and our customers are part of that same society. So, actually, we share the same basis for our strategies. Where we need more intense discussion is about the speed of transformation. Sometimes, industry wants to move ahead. Sometimes politicians want to move faster. I think speed is everything and the right acceleration. Let me take you through one example. For the transport sector, electromobility is the dominant approach to bringing down CO_2: The car's emissions directly factor into fleet regulation for new cars. Yet, acceptance of new electric vehicles by new car buyers and thereby the impact on the emissions of the transport sector are directly linked to the charging infrastructure. Without this ramping up fast, the decision to go for 0 g/km in new cars by 2035 could lead to lower new car sales and, consequently, to further ageing of the EU car fleet—the opposite of what is needed. That is why we have strongly urged to discuss how quickly the charging infrastructure in 27 European states can increase realistically and how one can set a target on the emissions of cars while at the same time agreeing on the path of charging infrastructure growth in the European Union. Yet, today, we still see a strong fragmentation of the EU markets: where infrastructure availability is highest, so is the BEV share—and vice versa. Therefore, we need a very thorough evaluation of the relation between supply-side regulation and demand-side prerequisites when the decision to go for 0 g/km in 2035 will be re-evaluated in 2026.

How Does Charging Infrastructure for Electromobility Affect Target Setting?

Grimm: It seems to me that one of the challenges is that the charging infrastructure is not predictable for the future. As a consequence, people do not know whether they can drive a battery electric vehicle everywhere they want to go. That, of course, will affect their choice to buy it, and that will affect your choice as a car maker to produce them.

Zipse: Well, we are aware of the infrastructure challenges, which is why we offer our customers the power of choice between different drivetrains.

Grimm: Basically, the same situation occurs in the field of hydrogen mobility. Especially in heavy-duty mobility, there has to be an infrastructure to initiate investments in the production of cars and heavy-duty cars. This is a big problem. Do you think politics acts in the right way here? What kind of strategy could one have, maybe to scale infrastructure up in a foreseeable way and also to co-finance it. It does not have to be the case that the state provides all the funding, but maybe there are interested investors who can put their money into it, because it could be a good business model. As soon as heavy-duty mobility based on hydrogen accelerates, a lot of money could be invested.

Zipse: We are well-advised to make the first step, because politicians will rarely make a first step if they do not have the feeling that there is an industry that takes on the challenge. For instance, we introduced the first electric car at BMW in 2013—a long time ago. The decision to produce that car has already been made in 2009. We did that far before politicians even talked about it. Thereby, you provide the opportunity of being more progressive on the political side.

Grimm: Will you pursue a similar approach with hydrogen mobility?

Zipse: When it comes to hydrogen, we are now at the same point that we have been with electromobility 10 years ago. We show that it is possible to propel cars with a fuel cell. Then, of course, the infrastructure is being built. It is not always a chicken and egg problem. It is more like ping-pong, where everyone moves forward step by step, and we want to be part of that game. We want to prove that technologically, there is a lot more possible than the average person knows. It is our task to convince politicians to be sometimes more progressive and sometimes to look at the timeframe again to slow down a little bit.

How Does International Competition Shape the Market for Zero-Emission Mobility?

Grimm: There is also very fierce international competition, especially in the area of fuel cell cars. East Asia is already having them in serial production. That could be a challenge for Europe. At the same time, society often cannot imagine that there is a lot of scope for further technological development. If people cannot imagine how different things will be, it is difficult to make decisions in politics to set the stage, for example, as in the case of hydrogen mobility. There is a huge ongoing debate in Germany about it. However, I understand that you are not saying the market has settled on battery electric vehicles, but you state that there are plenty of opportunities for both—hydrogen vehicles and battery electric vehicles.

Becker: Let's take the electric vehicle market as an example. If we look at the situation in the electric vehicle market today, this development was preceded by a political decision, the California Zero-Emission Vehicle Mandate, which included the obligation to put zero-emission cars on the road. This policy choice tied the automotive industry to a binding and increasing quota of zero-emission vehicles, be it fuel cell or battery electric. A similar approach could have been taken towards quotas of hydrogen—not at the vehicle side but at the infrastructure side. Hydrogen, for example, would then have had to be used in proportion to mineral oil production. We would have a different automotive industry today had such a choice been made. It is all about policy choices.

Grimm: So, on the one hand, it is the industry that innovates and, on the other, the regulatory dimension that triggers the strategic decision to bring a particular technology to the mass market.

Becker: As you rightly said, when you look at the demand side of fuel cells, it is an infrastructure choice that has to be made. It is about a concerted effort by industry and governments, by those who produce the energy, and by those who convert it into hydrogen. Therefore, our message is very clear in that respect. We feel very competent to supply the cars. We are developing different technologies to provide a space for politicians to make decisions, but ultimately it is their decision. In this regard, we have a very fragmented situation globally. If you look at Korea, they have a clear hydrogen agenda. If you look at California, there is no car that gets more credits than a fuel cell vehicle. The race is on. Sometimes, in Germany, we tend to overlook the fact that we are just one market and that people think differently in different places.

How Does Consumer Behaviour Influence Decision-Making Towards Sustainability?

Grimm: I think another big challenge is that it is very difficult for people to imagine how much will change in the future. It could be resources; as you mentioned: you need completely different resources and completely different supply chains, which will affect energy trade. Therefore, we have to move from importing fossil fuels to importing renewables, and these are big changes that have to become a reality quite rapidly. One of my questions would be: To what extent do you think that your customers already see what needs to change? At the same time, there is this continuing debate about the price of petrol, which is somewhat counterintuitive, because of course fossil fuels have to become more expensive if we want to become more sustainable. Something has to happen to make it more attractive to have a zero-emission car than a car that runs on fossil fuels. Hence, the other question is: How do we proceed in that direction in a situation where people, once they are affected, also react quite reluctantly?

Zipse: Well, the customer is the big unknown in this equation because a customer's behaviour is different from his/her political behaviour. What the consumer decides to vote for in terms of a political party and what the consumer decides to buy can be very different. Hence, basing one's product strategy on political elections is a very dangerous thing to do. People buy cars for many reasons, emotional reasons, cost reasons, and factual reasons, and I think you always have to observe and detect changes in customer behaviour. Customers do not change their behaviour overnight. Electromobility is a good example to support my point of view. It has taken more than 10–15 years, and there is still a lot of mistrust: Where are the charging points? How much will it cost me? Does the battery last long enough? Range anxiety; you have all that.

Grimm: So, what could be a solution to the uncertainties surrounding new technologies in the marketplace?

Zipse: A good piece of advice again is to look at the customer. We think there is room for a hydrogen market, namely for larger cars and especially in the premium sector. There are always first movers who want to have the latest technology. I think a premium brand like BMW suits progressive new technologies. It does not have to be scalable right away, but there is a new, growing market that could be very interesting for us.

Is the International Infrastructure Ready for New Sustainable Vehicles?

Grimm: We don't just have to think about what we do in Germany. In Germany, we have to think about what the world does and how we can contribute to it. How do you see the infrastructure issue? As far as hydrogen supply is concerned, do you see the possibility of scaling it up for private cars using the heavy-duty vehicle infrastructure, or do you think a different infrastructure is needed? Are there synergy effects?

Becker: If you provide hydrogen at the petrol station, would you really tell people that they have to have three axles or four in order to qualify for the use of hydrogen? I mean, if you provide the infrastructure, you should let everybody use it whenever there is a demand.

Zipse: I think there is an important difference with electromobility. Electromobility is geared towards privately owned cars. Hydrogen, on the other hand, is more of a cross-sector phenomenon in the energy market at the moment. There is not going to be a hydrogen infrastructure built just for private cars. It will probably start with steel production and then be used in the chemical industry and, regarding transportation, particularly in the heavy-duty sector. Eventually, individual transport may be able to use this infrastructure in the future. That is why there is an opportunity to develop the industrial infrastructure and the transport infrastructure for trucks and personal vehicles at the same time. This development is in contrast to electromobility, which started with small cars.

Grimm: I share the opinion that it does make sense to have this dual use of infrastructure. Very often, the opponents of it argue that it is wasteful, basically because you have two energy infrastructures for mobility. On the other hand, you have this heavy-duty mobility, where you need something other than battery electric vehicles anyway. So, you can also take advantage of this situation.

Zipse: It is possible that the structures will not be built in parallel. In the transport sector, hydrogen infrastructure will be built specifically for those cases where, for some reason, you cannot have electric mobility.

What I found compelling about hydrogen is the fact that we are already developing infrastructure for this kind of cases: we are currently using hydrogen mainly inside our buildings and factories, for example in the Leipzig plant, but also in the United States, where we use hydrogen for the internal logistics system.

Grimm: Why is that?

Zipse: If you want to become emission-free, the easiest way to do it is within closed systems, so we already have some experience in that area and it works really well for us. We are not just at the beginning of this technology. Hydrogen is one of the few elements that can be used with or without methane as an energy storage medium, so it can substitute or be combined with natural gas—which you cannot easily do with electromobility. This is the point at which individual mobility may choose the fuel cell as a drive system, particularly where you do not have access to a charging point or where hydrogen is set as a substitute for methane and where it will be more efficient to use it directly than generating electricity from hydrogen. That will be the case in many places, such as Japan. They will not have the same extensive charging infrastructure that we can build up here in Germany.

What Is the Status of Hydrogen Infrastructure Development?

Zipse: Electromobility is currently on the political agenda. There is an awareness that an infrastructure for electric mobility is necessary. Every country, be it France, Germany or even the European Union, is going to invest a lot of money in charging infrastructure. It is much more difficult to predict political behaviour on the hydrogen side. How much is the hydrogen infrastructure lagging behind the development of the electromobility infrastructure?

Grimm: I think it depends very much on the political decisions that are taken now. In the case of battery electric vehicles, there are the first vehicles, and now there is pressure to expand the infrastructure. With hydrogen development, it is more simultaneous. Regarding hydrogen, we are talking about cars and vehicles, and at the same time, there is a debate about infrastructure.

For heavy-duty mobility, there is still a debate about whether you can do it with battery electric vehicles, which I doubt. There are many problems involved if you want to realise long-distance transport. It is very effective to scale up hydrogen mobility by implementing or scaling up the logistics first. I see that there is an ongoing debate on prioritising and first using hydrogen in heavy industry, where it is needed in the chemical industry. And I think there's a big group of people who argue very much that battery electric vehicles will do it in mobility, which I doubt.

Zipse: Right. Do you think, due to the multi-use of hydrogen, we will ever be able to scale it up quickly enough? The chemical industry, steel industry, truck industry, car industry—everyone wants green hydrogen and it all must be green and not blue, not grey, and not brown, you know?

Grimm: In Germany, everything must be green immediately. In other countries, for example, in East Asia, they are much more pragmatic. They are scaling up the technology by being pragmatic about the 'colour' of hydrogen. I think, also in Europe, many countries are working with the so-called red hydrogen from nuclear power or with hydrogen produced from the country's energy mix and are trying to scale up the technology. I think, in Germany, we should be a little bit more pragmatic in this transition period until green hydrogen is available on a larger scale. At the same time, this would allow us to have a lot more hydrogen available during this transition period. Then, of course, we can hope that green hydrogen will become competitive in the long term, comparable to blue hydrogen, and then it would not be in short supply. But that is still to be decided, of course.

Becker: One factor that I think is closely related to the points you made is the organisation of a competitive market for decarbonisation. We strongly believe that emissions trading is a key element of sustainability. Today, when I charge my car, I use electricity that has been produced under the conditions of emissions trading with a cap. If I use diesel, that is not the case. If we make that car or the aluminium for that car in Landshut, it is part of the emissions trading system. So, wouldn't it make sense to extend this approach to transport? No matter what the energy source, be it hydrogen, liquid fuels, or electricity, it should all be priced according to CO_2 content. Then, you would see the performance in the price.

Grimm: You have said something very important. If you charge a battery electric vehicle, then it is—to a certain extent—carbon-free because there is a cap on the emissions in the trading system. So, it does not matter how carbon-intensive the energy mix is, but if you think about mobility in terms of the emission trading system, carbon intensity plays an important role. I think it would be wise to extend it.

Becker: What do you think would be a regulatory solution that would allow for technological openness?

Grimm: It is a very hard political battle that has to be fought, of course, at the level of the European Union. First of all, to include all the sectors and then also to include all the countries, because different countries have very different conditions. Some rely heavily on carbon-intensive energy production, others on nuclear power, which is also problematic from a German perspective. I think it is worth moving in this direction. Setting a carbon price is not the only tool. I think a carbon price is very predictable in the sense that if you know that emissions are capped and the cap decreases year by year. You can live with that to a certain extent. It is predictable and therefore a typical business risk that we are used to. Moreover, fossil fuel prices are fluctuating as well and you have to deal with the consequences for the economy. So, above all, it is very difficult to predict what particular regulation different governments will implement in the future.

How Can Carbon-Pricing Mechanisms Be Developed on a European Level?

Zipse: Designing market mechanisms that correlate sustainability goals with the actual behaviour of all participants would be a first step towards a Green Deal for Europe. That is not happening at the moment. Currently, each industry is assigned different sustainability targets. It would be the perfect time to introduce a price on CO_2, which, of course, would give you a big lever on achieving the targets. Why is this not happening?

Grimm: Economists have been proposing this concept for years. It is interesting that it is now being discussed so much in politics. The negotiation process at the European level makes the setting of targets difficult because some countries would be more affected than others. Therefore, you would have to compensate at the European Union level. In addition, this measure is unpopular—as you can see from the debate on petrol prices—because it visibly increases prices. A socially acceptable way must be found to reduce the burden on the customers. In other respects, this debate is not loud enough because it would be easy, for example, to increase CO_2 prices and at the same time reduce the price of electricity, which is currently dominated by taxes and levies. Therefore, you could very easily reduce it by a third, which would cost a typical household 400 euros less per year. There are possibilities, but it is very difficult to discuss these matters. Of course, there are different groups in society with different interests.

Becker: Looking at fuels, it shows that a trading system could provide a visible incentive for producers to decarbonise their products. For example, blending hydrogen-based synfuels into petrol could keep products affordable for customers. This example could then support a political debate—but that seems to be very controversial at the moment.

How Can Private Investments Support a European Green Deal?

Grimm: I think it is relatively easy to agree on targets, for example, climate neutrality by 2050 or 2045. However, it is very difficult to agree on the path, the mechanisms, and the regulation that will get us there. There are different perspectives on this issue. One part of the discussion focuses on the CO_2 price and market-oriented measures. Another part of the discussion emphasises that the state has to play a much more important role, that public spending has to be increased a lot in order to achieve climate neutrality. This is a misconception because investments in the private sector are just as necessary. Private investments account for 85% of total investments in Germany. So, in total, there is only 15% public investment compared to 85% private investment. However, increasing public investment will be difficult to steer in the right direction because of the fierce debate going on about what is the right way forward. Which sectors should be decarbonised first?

Zipse: And what is your opinion on that?

Grimm: There is a lot of dispute and no agreement on how to proceed. I would think that we need to agree on measures that will enable companies to invest. That means, on the one hand, the price of CO_2, of course, and the expansion of infrastructure in a predictable way. On the other hand, we also need more venture capital, and we need more capital in order to scale up innovation and to really establish the production of new climate technologies and applications in Europe. At the moment, we have much more venture capital opportunities in the US and even in East Asia. I would like to hear your thoughts on how to increase venture capital opportunities in Europe?

Zipse: I think, as you said before, we need to increase private investment in infrastructure in Europe. We need to consider a joint venture between policy makers, the banking sector, and the private sector to form consortia. This is the only way to really share the risks involved in infrastructure investments. Maybe that is the European route. I think Europeans are willing to take risks if the burden is shared. However, we have to take more risks in the future. That is for sure.

2.6 Science-Based Targets: Opportunities and Challenges of Setting Emissions Targets at the Company Level

The preceding expert conversation underscored the interconnectedness of national policies, market regulation, and corporate actions in navigating the Road to Net Zero. Achieving alignment with the evolving climate science and policy necessitates that corporations not only comprehend what is expected of them, but that they also dispose of robust methodologies to evaluate—and where needed, enhance—their mitigation strategies. With this in mind, this section turns its attention towards the concept of Science-Based Targets (SBTs) for ambitious GHG emission reductions. This is even more relevant, as the EU will require companies from 2024 onwards to report on their strategy for achieving compatibility with the 1.5° target using 'science-based' methods.

As underscored in Sect. 2.3, global climate policy has fostered a collective commitment to the Net Zero goal and the aim to limit global warming to well below 2 °C. This commitment requires ambitious emissions reduction trajectories that align with the 1.5 or 2 °C goal, respectively. SBTs provide a framework to translate these broad reduction pathways into tangible, corporate-level action plans. Leveraging insights from climate science, corporations can discern the specific degree of decarbonisation necessary for their unique context.

The objective of the SBTs is to offer clear directives for corporate climate action and enhance transparency regarding the alignment of emission reductions with the Paris Agreement. However, the methodology for establishing SBTs is multifaceted and continuously evolving. The goal of this review is not to delve deeply into these complexities, but rather to furnish an overview of the overall rationale and highlight pertinent considerations for its application, critical assessment, and future development.

2.6.1 The Science-Based Target Initiative: Origin and Mission

Around the time when the Paris Agreement was adopted, the CDP (Carbon Disclosure Project), the United Nations Global Compact, the World Resources Institute, and the World Wildlife Fund formed the Science-Based Target Initiative (SBTi) to develop a standard to derive GHG reduction targets that are aligned with the 2 °C or, respectively, with the well below 2 °C temperature goal of Paris at the company level (Bjørn et al., 2022). More recently, the SBTi increased its ambition level to focus on the 1.5 °C target.

Subsequently, the fundamental premise of SBTs at the individual (corporate) level has evolved into a pragmatic, data-informed, target-setting method and validation under the SBTi enabling businesses to align their strategies with the Paris Agreement's objectives. In 2021, more than 2000 companies from 70 countries, accounting for 35% of global market capitalisation according to the SBTi Progress Report, have either committed to setting SBTs or have already had their SBTs approved (SBTi, 2022). This number continues to grow, with more than 4000 companies setting targets by the end of 2022 (SBTi, 2023a).

The mission of the SBTi is to drive 'ambitious climate action in the private sector by enabling organizations to set science-based emissions reduction targets' (SBTi, 2023a). The SBTi seeks to accomplish this mission by providing a framework and guidelines for businesses to set and validate their GHG reduction targets. Facilitating a process to independently assess and validate companies' targets, the SBTi provides an external assessment of corporate emission reduction targets and promotes transparency by publicly recognising companies that have set science-based targets. What the SBTi does not do (and does not intend to do) is to verify the reported data and actual business performance.

As a multi-stakeholder initiative, SBTi's funding relies on target validation fees and contributions from various corporate and charitable entities. Given its standing as a private non-profit organisation with substantial global influence, there has been a discussion regarding the absence of a public entity or policy to carry out the functions of the initiative (see Bjørn et al., 2022; Lister, 2018; Marland et al., 2015; Trexler & Schendler, 2015). Despite facing initial criticism (cf. Trexler & Schendler, 2015), the SBTi has nevertheless evolved as the globally most acknowledged framework for setting emission reduction targets.

2.6.2 The Science Base of the SBTs

The SBTi aims to mobilise 'the private sector to take the lead on urgent climate action' by 'enabling organizations to set science-based emissions reduction targets' (SBTi, 2023a). But what exactly constitutes the 'science' in Science-Based Targets?

The labelling of targets in the political and business arena as 'Science-Based' might initially seem contradictory, as 'operational targets are socio-political choices' (Andersen et al., 2021, p. 2). While climate science can delineate the phenomena, causes, and repercussions of global warming, the decision to halt or control climate change involves normative judgements that go beyond the scope of science. Accordingly, the IPCC's 4th assessment report (IPCC, 2007, p. 64) stressed that defining what constitutes a 'dangerous anthropogenic interference with the climate system' is only partially rooted in science 'as it inherently involves normative judgements'.

Against this background, the Paris goals of limiting global warming are 'science-based' in the context of being scientifically informed but politically determined. Climate science asserts that the hazards of global warming rise sharply beyond 1.5 and 2 °C. However, the aspiration to avert these risks, as codified in the Paris Agreement, is a value-based choice informed by scientific findings, but ultimately decided politically.

In crafting science-based targets for corporations, the SBTi takes the political commitment to the Paris Agreement's goals as its starting point. Under this framework, corporate emission reduction objectives are deemed 'science-based' if they align with 'what the latest climate science says is necessary to meet the goals of the Paris Agreement' (SBTi, 2023c, p. 5). The 'science' in these targets maps the path required to achieve the globally accepted Paris objectives. This necessitates that the SBTs be quantifiable, measurable (Andersen et al., 2021) and guided by a methodology anchored in the emission reductions that climate science prescribes to fulfil the Paris goals. Importantly, as these targets are linked to the 'latest climate science', advances in climate science, such as revised estimates of the remaining carbon budget, could necessitate updates to SBTs and amplify the level of ambition required for climate action.

2.6.3 Which Emissions Count? Clarifying the Scope and Base Year for SBTs

Before explaining the different types of SBTs and how to calculate them, it is important to clarify the scope of emissions that companies need to consider according to the SBTi. Here, the SBTi leans on the carbon accounting methodology defined in the GHG protocol (see World Resources Institute [WRI] and World Business Council for Sustainable Development [WBCSD], 2004). This protocol categorises Scope 1 emissions as direct GHG emissions 'from sources that are owned or controlled by the company' (GHG Protocol, see WRI and WBCSD, 2004, p. 25), such as from combustion, production, or chemical processes. Scope 2 refers to indirect GHG emissions caused by energy consumption (including electricity, steam, heating and cooling energy), and Scope 3 refers to GHG emissions caused by activities neither controlled nor owned by the company. Scope 3 thus includes emissions that occur in a company's value chain, both upstream and downstream (GHG Protocol, see WRI and WBCSD, 2004, p. 25). In the case of automotive original equipment manufacturers (OEMs), this can include emissions from battery production (upstream) or vehicle emissions from customer cars (downstream).

In setting SBTs as per the latest guidelines, companies are required to cover a minimum of 95% of their Scope 1 and Scope 2 emissions (SBTi, 2023b). The situation with Scope 3 emissions is more intricate, as their obligatory inclusion relies on various other factors, which will be elaborated subsequently.

2.6.4 Different Types of Science-Based Targets

As the Road to Net Zero is a marathon and not a sprint, companies need targets that allow them to plan for both the immediate next steps and the long-term journey. Accordingly, the SBTi provides different types of SBTs.

Near-term targets focus on rapid and deep emission reductions that cover a minimum of 5 years and a maximum of 10 years from the date the target is submitted for validation. Near-term targets must be aligned with a 1.5 °C scenario. For most companies, this implies halving emissions by 2030. In addition to Scope 1 and 2 emissions, near-term emissions must cover at least 67% of all Scope 3 emissions if these indirect emissions account for more than 40% of a company's life cycle GHG inventory. For many companies, this is the case. To illustrate, after significant reductions of its Scope 1 and 2 emissions, BMW's Scope 3 emissions account for much more than 90% of its total emissions. Companies that have less than 40% Scope 3 emissions are encouraged to include them voluntarily.

Long-term targets indicate the degree of emission reductions that companies need to achieve by 2050 or sooner. Long-term targets must cover a minimum of 30 years from the date the target is submitted for validation and must be aligned with a 1.5 °C scenario, which means that most sectors must achieve at least a 90% reduction in absolute emissions by 2050 (or 2040 for the power sector) compared to a base year. For long-term targets, companies must cover at least 95% of their Scope 1 and 2 emissions and at least 90% of their Scope 3 emissions (irrespective of their relative share).

The net zero standard is a newer benchmark established by the SBTi and provides guidance to substantiate the path for a company to reach a science-based 'net zero' status. Apart from committing to long-term and short-term reductions (90%), companies are obligated to neutralise any remaining emissions. This entails utilising permanent carbon removals and storage methods (including nature-based solutions, such as restoring forests, soils, and wetlands, or technical solutions, such as direct-air capture) to counterbalance the final <10% of residual emissions that cannot be eliminated. Note that offsetting emissions through compensation measures that merely avoid emissions elsewhere (e.g. more efficient cook-stoves) are deemed insufficient (SBTi, 2023d).

2.6.5 Different Methods and Sector Approaches for Determining Necessary Reduction Levels

The crux of the SBTi, albeit intricate and somewhat complex, lies in the methodologies used to determine the exact emission reductions required to attain a specific ambition level. Previously, the SBTi offered calculation tools based on accepted target-setting methods from various sources, including public organisations, companies, and academia, resulting in a total of seven methods (Bjørn et al., 2021). More recently, however, the SBTi has refined its recommendations for Scope 1 and 2 emissions targets to two methods, focusing on either absolute reductions (company-wide) or relative reductions (emission intensities, such as per ton of product or per dollar of revenue) (Bjørn et al., 2022, p. 55).

Regardless of the method, the underlying approach considers the extent of emission reductions needed by all companies to achieve a particular temperature goal. In the SBTi's early framework versions, companies could align their ambition level with a 2 °C or well below 2 °C trajectory. However, given the recent warnings from climate science (cf. IPCC, 2018), the SBTi has raised its framework's ambition to aim for the 1.5 °C goal. Consequently, whereas the 2 °C or well below 2 °C pathway-aligned SBTs were previously approved, since mid-2022, the SBTi only approves SBTs consistent with the 1.5 °C pathway (SBTi, 2023b, p. 4).

With the new ambition level of 1.5 °C in mind, let us come back to the absolute and relative reduction approach. The former approach focuses on an absolute decrease in emissions, irrespective of a company's size, production output, or revenue. Known formerly as the 'absolute contraction approach' (ACA), this method applies broadly across sectors (with exceptions like agriculture) and demands that absolute emissions decrease by an amount at least consistent with the cross-sector pathway. For the 1.5 °C goals, this equates to most sectors reducing their Scope 1 and 2 emissions by a minimum of 4.2% annually (SBTi, 2023b, p. 3).

Contrarily, relative or 'intensity' approaches set emission reduction targets relative to a specific business metric, such as per unit of production (physical intensity) or per revenue unit (economic intensity). This approach enables companies to decrease greenhouse gas emission intensity while accommodating business growth. The SBTi employs corresponding pathways to model the necessary emission intensity reductions to align with the 1.5 °C goal, generally applicable for Scope 3 emissions.

Nevertheless, emission intensity and the scope for its improvement vary by sector, as do absolute emissions and their potential for reduction. Decarbonising power generation through a shift from coal to gas or solar, for instance, is simpler than decarbonising aviation, which relies on kerosene. Acknowledging the diverse mitigation opportunities and challenges that various sectors face, SBTi methodologies incorporate the Sectoral Decarbonisation Approach (SDA), using sector-specific emission scenarios from the International Energy Agency (IEA). Applied to the absolute reduction approach, the sector-specific reduction approach prescribes absolute reduction targets for specific sectors like agriculture or iron and steel. Similarly, the sector-specific intensity convergence approach applies relative reduction logic to sector-specific pathways. For the food, land use and agriculture sector (FLAG), there are even commodity-specific reduction pathways that model the necessary emission intensity reductions for commodities like beef, rice, leather, or dairy to align with the 1.5 °C goal (SBTi, 2023b, p. 6). The EU's CSRD reporting requirements foresee targets to be set in absolute terms. So, from 2024 onwards, the choice for European companies is effectively limited.

This overview underscores the multitude of methods for setting an SBT, creating significant complexity. In response, the SBTi provides small and medium-sized firms (SMEs) with a simplified procedure that offers flexibility. Even large corporations have some leeway in choosing their base year, target format (near-term or long-term), scope (optional for Scope 3 emissions) and, most importantly, the methods to define their target. Some academics see this diversity of methods as a strength, contending that 'there is not a single SBT method that is best in all sectors and company situations' (Aden, 2018, p. 1095). However, others criticise the potential for lack of comparability and for companies to choose less challenging targets (Bjørn et al., 2021; Freiberg et al., 2021). Against this backdrop, the SBTi continuously refines its methodology and provides more sector-specific guidance, yet questions persist about potential future refinements to the methodology.

2.6.6 Benefits and Challenges of Science-Based Targets

So far, setting SBTs has been voluntary. While the SBTi predicts widespread adoption, citing innovation theory that rapid diffusion occurs when a critical mass of early adopters is reached (SBTi, 2022), it remains to be seen whether the majority of companies will set SBTs without legislative pressure.

Independent of the further development and dissemination of the SBTi framework, research indicates two areas of potential positive impact from setting SBTs: corporate climate action and regulatory alignment, as well as public policy.

Regarding corporate climate action on the Road to Net Zero, the SBTi offers direction to companies, and through its validation, potentially encourages more ambitious corporate climate action regarding GHG reduction. For strategic direction, companies need operational targets for their planning because what gets measured gets done (cf. Chap. 1). This is particularly important when considering long-term strategic investments. For instance, in the automotive industry, the development and implementation of a new car platform with a projected lifetime of more than 20 years can be considered. To prepare adequately for the future, companies find it valuable to anchor their strategies in precise assumptions (cf. Chap. 3).

When it comes to potential incentives for more ambitious climate action and actual emission reductions by companies, the existing research is not conclusive. Initial studies, such as those by Freiberg et al. (2021) and Bolton and Kacperczyk (2023), have explored the impact of the SBTi on emissions reductions, but the results are largely inconclusive for making generalisable claims due to various influencing factors affecting targets and a company's climate ambition (Bjørn et al., 2022, p. 61). Nonetheless, companies that report SBTs for emission reduction appear to invest more than do companies that merely set internal targets (cf. Bjørn et al., 2022; Freiberg et al., 2021).

In terms of potential business benefits, science-based targets equip companies to better align with not only the present but also future regulatory environments. As discussed in Sect. 2.4, policymakers are currently implementing various policy instruments at the national level, creating a fragmented regulatory landscape for companies operating across borders. While it remains uncertain whether and how these diverse regulatory approaches will converge, it is evident that the Paris Agreement and its reduction targets will persist as a global benchmark guiding future implementation. By aligning their operations with this global goal, science-based targets can provide a universally relevant framework amid a disjointed regulatory landscape, thus reducing regulatory risk and uncertainty.

Regarding the evolution of public climate policy, both research and recent policy developments indicate a potentially positive impact of SBTs on public policy. Whereas early critics, such as Trexler and Schendler (2015, p. 933), argue that SBTs 'will only further delay policy', Marland et al. (2015) disagree with the claim that SBTs are just another distraction from solving the real problem and point to the importance of bottom-up initiatives in a democracy as a means of developing regulations from the dialogue that emerges. Especially in the absence of public policy, Banda (2018, p. 387) argues that 'private climate governance could help embed rules of public international law in the domestic sphere and drive up State ambition over time'. Similarly, recent policy developments in corporate sustainability reporting, such as the CSRD, illustrate how the voluntary GRI reporting framework and other voluntary standards have evolved over two decades into a new policy (cf. Chap. 4).

Therefore, while it is clear that private sector initiatives can stimulate public policy development, these frameworks alone cannot meet the challenge of achieving the 1.5 °C target without corresponding regulations and instruments, as discussed in Sect. 2.4. Furthermore, critics contend that the SBTi requires financial contributions from companies for registration and consultation, questioning the monetary independence of the initiative. Despite these criticisms, no other current framework that translates global climate goals to the corporate level has gained equal recognition or refinement.

The ongoing critiques of the SBTi underscore the fact that the SBT framework is still maturing and would benefit from further evolution and refinement. A key aspect of the SBTi's future progress involves enhancing the target-setting methodology. As of 2023, sector-specific guidance remains absent for several sectors, including iron and steel, chemicals, and oil & gas. Even with the guidance available for other sectors, the multitude of methods can result in potential misuse and a lack of transparency. Moreover, while SBTs differentiate between near-term (2030) and long-term (2050) targets, discussion is ongoing regarding the establishment of medium-term targets, especially concerning the necessary emissions reductions between 2030 and 2050. Ultimately, future enhancements are required to strengthen the comparability, guidance, and transparency of SBTs.

Finally, any methodology that sets reduction targets for carbon emissions will depend on the quality of the underlying GHG data and accounting. Currently, the SBTi leans on the carbon accounting methodology defined in the GHG protocol. Although widely used as the de facto global standard for carbon accounting, this methodology has several weaknesses. In particular, Scope 3 emissions pose significant challenges. Their measurement, due to the complexity of global supply chains, can be fraught with errors and potential bias.

The Protocol's allowance for using secondary or averaged data instead of specific emissions data (GHG Protocol, see WRI and WBCSD, 2004) often opens the door for the evasion or manipulation of Scope 3 measurements. Consequently, companies may strategically choose what to report. For instance, if a company has a supplier whose actual emissions greatly exceed the industry average, the company can reduce its reported emissions by using that averaged data. On the other hand, if a company invests in improving supplier performance, thereby lowering the industry average, competitors can exploit these industry averages to report emission reductions without making substantial changes to their own processes. This level of flexibility not only weakens the integrity of Scope 3 measurements but also does little to spur sincere decarbonisation efforts (Kaplan & Ramanna, 2021).

Consequently, the effectiveness of SBTs may be curtailed by these inherent limitations in carbon accounting as long as they continue to align with the GHG Protocol. Therefore, future improvements to the SBT framework should not solely focus on refining target-setting methods, but should also explore innovative approaches in carbon accounting. This could potentially involve adopting systems such as the proposed E-liability accounting system, which advocates for the use of inventory and cost accounting practices to accurately measure GHG emissions across corporate supply chains (Kaplan & Ramanna, 2022).

2.7 Conclusion

The Road to Net Zero symbolises a collective expedition requiring diverse entities to contribute their unique inputs, with a shared destination in sight. This chapter's objective was to lay the groundwork for this journey, detailing the climate science and policy contexts underpinning the Road to Net Zero, ahead of subsequent chapters that delve into the specific roles of and actions taken by corporations.

Climate science provides the essential scientific underpinnings (Sect. 2.2), while global climate policy, as exemplified by the Paris Agreement, fosters a collective commitment to limit global warming to well below 2 °C and target net zero emissions (Sect. 2.3). National policymakers have various tools at their disposal to establish and execute national emission reduction goals (Sects. 2.4 and 2.5), and for corporations, Science-Based Targets (SBTs) offer a framework to synchronise their climate initiatives with global policy objectives (Sect. 2.6).

Reflecting upon this foundational chapter, we highlight five significant takeaways that encourage further discourse:

1. To cap global warming at 2 °C, or ideally at 1.5 °C, thereby averting perilous climate change, swift and considerable emissions reductions are imperative. This indicates that unless swift decarbonisation transpires, the remaining carbon budget will soon be depleted.
2. The Paris Agreement's legal commitment to a precise temperature target and the objective of achieving net zero emissions fundamentally necessitates a comprehensive economic transformation. Unlike Kyoto's incremental reductions, the Net Zero target demands a profound transition from our current fossil fuel-based economy to extensive decarbonisation.
3. National climate policies are instrumental in facilitating emission reductions through bolstering market-based solutions, augmented by supportive schemes and standards setting where needed. This policy mix needs to negotiate a balance between ecological efficacy, fair contribution to global reductions, economic efficiency, political and administrative feasibility, and impacts on competitiveness and innovation.
4. For long-term decarbonisation, public policies need to offer certainty and unambiguous guidance on targets. Promoting technological openness in achieving these targets can stimulate innovation, enhance resilience, and boost efficiency. Moreover, infrastructure is key to decarbonising the transport sector. Policy plays a crucial role in collaboration with corporate players to ramp up the availability of charging for battery electric vehicles, as well as for hydrogen-powered fuel cell electric trucks and cars.
5. SBTs furnish a framework enabling businesses to align their strategies with cutting-edge climate science and global public policy commitments. Despite their limitations, SBTs serve as crucial tools for augmenting the efficacy and credibility of business-driven decarbonisation. To better capitalise on this potential, future developments of SBTs will profit from further methodological refinement, improved carbon accounting, and enhanced legitimacy of the underlying standard-setting process.

The Road to Net Zero thus embodies a collective endeavour wherein companies play an indispensable role but it also necessitates an enabling environment fortified by public policies and robust science. The subsequent chapters of this book will explore the transformative changes that companies can engender in their business strategy, reporting, products, value chains, production, and technology. Starting this exploration, Chap. 3 *'Creating Corporate Sustainability Strategy. From Integrated Thinking to Integrated Management'* will discuss these ideas in greater detail. A novel and more integrated perspective is required to navigate the challenges of the Road to Net Zero and other sustainability issues.

References

Acemoglu, D., Aghion, P., Bursztyn, L., & Hemous, D. (2012). The environment and directed technical change. *American Economic Review, 102*(1), 131–166. https://doi.org/10.1257/aer.102.1.131

Aden, N. (2018). Necessary but not sufficient: the role of energy efficiency in industrial sector low-carbon transformation. *Energy Efficiency, 11*(5), 1083–1101. https://doi.org/10.1007/s12053-017-9570-z

Aldy, J. E., Krupnick, A. J., Newell, R. G., Parry, I. W. H., & Pizer, W. A. (2010). Designing climate mitigation policy. *Journal of Economic Literature, 48*(4), 903–934. https://doi.org/10.1257/jel.48.4.903

Andersen, I., Ishii, N., Brooks, T., Cummis, C., Fonseca, G., Hillers, A., Macfarlane, N., Nakicenovic, N., Moss, K., Rockström, J., Steer, A., Waughray, D., & Zimm, C. (2021). Defining 'science-based targets'. *National Science Review, 8*(7), nwaa186. https://doi.org/10.1093/nsr/nwaa186

Banda, M. L. (2018). The bottom-up alternative: the mitigation potential of private climate governance after the Paris Agreement. *Harvard Environmental Law Review, 42*, 325–389.

Barrett, S. (2005). Global climate change and the Kyoto Protocol. In S. Barrett (Ed.), *Environment and statecraft: The strategy of environmental treaty-making* (pp. 359–398). Oxford University Press. https://doi.org/10.1093/0199286094.003.0015

Baumol, W. J., & Oates, W. E. (1988). *The theory of environmental policy*. Cambridge University Press.

Bjørn, A., Lloyd, S., & Matthews, D. (2021). From the Paris Agreement to corporate climate commitments: evaluation of seven methods for setting 'science-based' emission targets. *Environmental Research Letters, 16*(5), 54019. https://doi.org/10.1088/1748-9326/abe57b

Bjørn, A., Tilsted, J. P., Addas, A., & Lloyd, S. M. (2022). Can science-based targets make the private sector Paris-aligned? A review of the emerging evidence. *Current Climate Change Reports, 8*(2), 53–69. https://doi.org/10.1007/s40641-022-00182-w

Bodansky, D. (2001). The history of the global climate change regime. In U. Luterbacher & D. F. Sprinz (Eds.), *International relations and global climate change* (pp. 23–40). MIT Press.

Bodansky, D. (2016). The Paris climate change agreement: A new hope? *American Journal of International Law, 110*(2), 288–319. https://doi.org/10.5305/amerjintelaw.110.2.0288

Bolton, P., & Kacperczyk, M. (2023). *Firm commitments* (National Bureau of Economic Research Working Paper Series, 31244). https://doi.org/10.3386/w31244

Bundesministerium der Finanzen. (2023). Kfz-Steuer-Rechner. Retrieved June 16, 2023, from https://www.bundesfinanzministerium.de/Web/DE/Service/Apps_Rechner/KfzRechner/KfzRechner.html

Bundesministerium für Digitales und Verkehr. (2021). BMDV fördert öffentliche Wasserstofftankstellen für Nutzfahrzeuge. Retrieved June 16, 2023, from https://bmdv.bund.de/SharedDocs/DE/Artikel/G/foerderung-oeffentliche-wasserstofftankstellen-nutzfahrzeuge.html?nn=12830

Bundesministerium für Wirtschaft und Klimaschutz. (2023). Erneuerbare-Energien-Gesetz. Retrieved June 16, 2023, from https://www.erneuerbare-energien.de/EE/Redaktion/DE/Dossier/eeg.html?docId=72b6752a-1ad5-4029-a969-1ddd04e938d9

Deutscher Bundestag. (2022a). Gesetz über einen nationalen Zertifikatehandel für Brennstoffemissionen: Brennstoffemissionshandelsgesetz.

Deutscher Bundestag. (2022b). Gesetz zur Reduzierung und zur Beendigung der Kohleverstromung: Kohleverstromungsbeendigungsgesetz.

Deutscher Bundestag. (2023). Gesetz für den Ausbau erneuerbarer Energien: Erneuerbare-Energien-Gesetz.

European Commission. (2023a). A European Green Deal: Striving to be the first climate-neutral continent. Retrieved June 15, 2023, from https://commission.europa.eu/strategy-and-policy/priorities-2019-2024/european-green-deal_en

European Commission. (2023b). Carbon Border Adjustment Mechanism. Retrieved July 6, 2023, from https://taxation-customs.ec.europa.eu/carbon-border-adjustment-mechanism_en

European Commission. (2023c). Circular economy action plan: The EU's new circular action plan paves the way for a cleaner and more competitive Europe. Retrieved June 16, 2023, from https://environment.ec.europa.eu/strategy/circular-economy-action-plan_en

European Commission. (2023d). Delivering the European Green Deal. Retrieved June 16, 2023, from https://commission.europa.eu/strategy-and-policy/priorities-2019-2024/european-green-deal/delivering-european-green-deal_en

European Commission. (2023e). EU Emissions Trading System (EU ETS). Retrieved June 16, 2023, from https://climate.ec.europa.eu/eu-action/eu-emissions-trading-system-eu-ets_en

European Commission. (2023f). European Climate Law. Retrieved June 16, 2023, from https://climate.ec.europa.eu/eu-action/european-green-deal/european-climate-law_en

European Parliament. (2022). Climate change: Deal on a more ambitious Emissions Trading System (ETS). Retrieved July 6, 2023, from https://www.europarl.europa.eu/news/en/press-room/20221212IPR64527/climate-change-deal-on-a-more-ambitious-emissions-trading-system-ets

European Parliament. (2023). Fit for 55: zero CO2 emissions for new cars and vans in 2035. Retrieved June 22, 2023, from https://www.europarl.europa.eu/news/en/press-room/20230210IPR74715/fit-for-55-zero-co2-emissions-for-new-cars-and-vans-in-2035

European Parliament, & European Council. (2019). Regulation (EU) 2019/631 of the European Parliament and of the Council of 17 April 2019 setting CO2 emission performance standards for new passenger cars and for new light commercial vehicles, and repealing Regulations (EC) No 443/2009 and (EU) No 510/2011 (recast) (Text with EEA relevance).

European Parliament, & European Council. (2021). Regulation (EU) 2021/1119 of the European Parliament and of the Council of 30 June 2021 establishing the framework for achieving climate neutrality and amending Regulations (EC) No 401/2009 and (EU) 2018/1999: European Climate Law.

Freiberg, D., Grewal, J., & Serafeim, G. (2021). Science-*based carbon emissions targets* (Harvard Business School Working Paper). Advance online publication. https://doi.org/10.2139/ssrn.3804530

Garnadt, N., Grimm, V., & Reuter, W. (2020). *Carbon adjustment mechanisms: Empirics, design and caveats* (German Council of Economic Experts Working Paper, 11/2020).

Goulder, L. H., & Parry, I. W. H. (2008). Instrument choice in environmental policy. *Review of Environmental Economics and Policy, 2*(2), 152–174. https://doi.org/10.1093/reep/ren005

Intergovernmental Panel on Climate Change (IPCC). (2007). Synthesis report. In Core Writing Team, R. Pachauri, & A. Reisinger (Eds.), *Climate change 2007: Synthesis report: Contribution of Working Groups I, II and III to the fourth assessment report of the Intergovernmental Panel on Climate Change* (pp. 23–74) IPCC.

Intergovernmental Panel on Climate Change (IPCC). (2014). Summary for policymakers. In Core Writing Team, R. Pachauri, & L. A. Meyer (Eds.), *Climate change 2014: Synthesis report: Contribution of Working Groups I, II and III to the fifth assessment report of the Intergovernmental Panel on Climate Change* (pp. 1–32) IPCC.

Intergovernmental Panel on Climate Change (IPCC). (2018). Summary for policymakers. In V. Masson-Delmotte, P. Zhai, H.-O. Pörtner, D. Roberts, J. Skea, P. R. Shukla, et al. (Eds.), *Global warming of 1.5 °C.: An IPCC special report on the impacts of global warming of 1.5°C above pre-industrial levels and related global greenhouse gas emission pathways, in the context of strengthening the global response to the threat of climate change, sustainable development, and efforts to eradicate poverty* (pp. 3–24). Cambridge University Press.

Intergovernmental Panel on Climate Change (IPCC). (2021a). IPCC factsheet. What is the IPCC? Retrieved June 16, 2023, from https://www.ipcc.ch/site/assets/uploads/2021/07/AR6_FS_What_is_IPCC.pdf

Intergovernmental Panel on Climate Change (IPCC). (2021b). IPCC factsheet. What literature does the IPCC assess? Retrieved June 16, 2023, from https://www.ipcc.ch/site/assets/uploads/2021/07/AR6_FS_assess_literature.pdf

Intergovernmental Panel on Climate Change (IPCC). (2021c). Summary for policymakers. In V. Masson-Delmotte, P. Zhai, A. Pirani, S. L. Connors, C. Péan, S. Berger, et al. (Eds.), *Climate change 2021: The physical science basis: Contribution of Working Group I to the sixth assessment report of the Intergovernmental Panel on Climate Change* (pp. 3–32). Cambridge University Press.

Intergovernmental Panel on Climate Change (IPCC). (2022). Summary for policymakers. In H.-O. Pörtner, D. C. Roberts, M. Tignor, E. S. Poloczanska, K. Mintenbeck, A. Alegría, et al. (Eds.), *Climate change 2022: Impacts, adaptation, and vulnerability: Contribution of Working Group II to the sixth assessment report of the Intergovernmental Panel on Climate Change* (pp. 3–33). Cambridge University Press.

Intergovernmental Panel on Climate Change (IPCC). (2023). Summary for policymakers. In Core Writing Team, H. Lee, & J. Romero (Eds.), *Climate change 2023: Synthesis report: A report of the Intergovernmental Panel on Climate Change. Contribution of Working Groups I, II and III to the sixth assessment report of the Intergovernmental Panel on Climate Change*.

Jaffe, A. B., Newell, R. G., & Stavins, R. N. (2002). Environmental policy and technological change. *Environmental and Resource Economics, 22*(1/2), 41–70. https://doi.org/10.1023/A:1015519401088

Kaplan, R. S., & Ramanna, K. (2021). Accounting for climate change. *Harvard Business Review, 99*(6), 120–131.

Kaplan, R. S., & Ramanna, K. (2022). We need better carbon accounting. Here's how to get there. *Harvard Business Review Digital Articles*.

Klinsky, S., Roberts, T., Huq, S., Okereke, C., Newell, P., Dauvergne, P., O'Brien, K., Schroeder, H., Tschakert, P., Clapp, J., Keck, M., Biermann, F., Liverman, D., Gupta, J., Rahman, A., Messner, D., Pellow, D., & Bauer, S. (2017). Why equity is fundamental in climate change policy research. *Global Environmental Change, 44*, 170–173. https://doi.org/10.1016/j.gloenvcha.2016.08.002

Lister, J. (2018). The policy role of corporate carbon management: Co-regulating ecological effectiveness. *Global Policy, 9*(4), 538–548. https://doi.org/10.1111/1758-5899.12618

Marland, G., Kowalczyk, T., & Cherry, T. L. (2015). "Green fluff"? The role of corporate sustainability initiatives in effective climate policy: Comment on "Science-based carbon targets for the corporate world: The ultimate sustainability commitment, or a costly distraction?". *Journal of Industrial Ecology, 19*(6), 934–936. https://doi.org/10.1111/jiec.12343

National Oceanic and Atmospheric Administration. (2022). Carbon dioxide now more than 50% higher than pre-industrial levels: News & Features. Retrieved June 16, 2023, from https://www.noaa.gov/news-release/carbon-dioxide-now-more-than-50-higher-than-pre-industrial-levels

Newell, R. G., & Stavins, R. N. (2003). Cost heterogeneity and the potential savings from market-based policies. *Journal of Regulatory Economics, 23*(1), 43–59. https://doi.org/10.1023/A:1021879330491

Nordhaus, W. (2019). Climate change: The ultimate challenge for economics. *The American Economic Review, 109*(6), 1991–2014.

Pigou, A. C. (1920). *The economics of welfare*. Macmillan.

Prentice, I. C., Baines, P. G., Scholze, M., & Wooster, M. J. (2012). Fundamentals of climate change science. In S. E. Cornell, I. C. Prentice, J. I. House, & C. J. Downy (Eds.), *Understanding the earth system: Global change science for application* (pp. 39–71). Cambridge University Press. https://doi.org/10.1017/CBO9780511921155.005

Presse- und Informationsamt der Bundesregierung. (2022). Neue Förderregeln für den Umweltbonus ab 2023. Retrieved June 16, 2023, from https://www.bundesregierung.de/breg-de/schwerpunkte/klimaschutz/eenergie-und-mobilitaet/faq-umweltbonus-1993830

Rogelj, J., den Elzen, M., Höhne, N., Fransen, T., Fekete, H., Winkler, H., Schaeffer, R., Sha, F., Riahi, K., & Meinshausen, M. (2016). Paris Agreement climate proposals need a boost to keep warming well below 2 °C. *Nature, 534*(7609), 631–639. https://doi.org/10.1038/nature18307

Sachverständigenrat zur Begutachtung der gesamtwirtschaftlichen Entwicklung. (2020). Corona-Krise gemeinsam bewältigen, Resilienz und Wachstum stärken: Jahresgutachten 20/21. Statistisches Bundesamt.

Science Based Targets Initiative (SBTi). (2022). Science-based net zero: Scaling urgent corporate climate action worldwide (science based targets initiative annual progress report, 2021). https://sciencebasedtargets.org/resources/files/SBTiProgressReport2021.pdf. Accessed 22 June 2023.

Science Based Targets Initiative (SBTi). (2023a). About us. Retrieved June 22, 2023, from https://sciencebasedtargets.org/about-us

Science Based Targets Initiative (SBTi). (2023b). Getting started guide for science-based target setting. https://sciencebasedtargets.org/resources/files/Getting-Started-Guide.pdf. Accessed 22 June 2023.

Science Based Targets Initiative (SBTi). (2023c). SBTi corporate manual: TVT-INF-002, Version 2.1. https://sciencebasedtargets.org/resources/files/SBTi-Corporate-Manual.pdf. Accessed 22 June 2023.

Science Based Targets Initiative (SBTi). (2023d). The corporate net-zero standard. Retrieved June 22, 2023, from https://sciencebasedtargets.org/net-zero

Shakun, J. D., Clark, P. U., He, F., Marcott, S. A., Mix, A. C., Liu, Z., Otto-Bliesner, B., Schmittner, A., & Bard, E. (2012). Global warming preceded by increasing carbon dioxide concentrations during the last deglaciation. *Nature, 484*(7392), 49–54. https://doi.org/10.1038/nature10915

Stavins, R. N. (2003). Chapter 9—Experience with market-based environmental policy instruments. In K.-G. Mäler & J. R. Vincent (Eds.), *Handbook of environmental economics: Environmental degradation and institutional responses* (pp. 355–435). Elsevier. https://doi.org/10.1016/S1574-0099(03)01014-3

Stern, N. H. (2007). *The economics of climate change: The Stern review*. Cambridge University Press.

Tietenberg, T. H. (2006). *Emissions trading: Principles and practice*. Resources for the Future.

Tietenberg, T. H., & Lewis, L. (2018). *Environmental and natural resource economics* (11th ed.). Routledge.

Trexler, M., & Schendler, A. (2015). Science-based carbon targets for the corporate world: The ultimate sustainability commitment, or a costly distraction? *Journal of Industrial Ecology, 19*(6), 931–933. https://doi.org/10.1111/jiec.12311

Umweltbundesamt (UBA). (2022). Wirkung des nationalen Brennstoffemissionshandels – Auswertungen und Analysen Grundlagen für den ersten Erfahrungsbericht der Bundesregierung gemäß § 23 BEHG im Jahr 2022. Retrieved September 27, 2023, from https://www.umweltbundesamt.de/sites/default/files/medien/1410/publikationen/2022-12-02_climate-change_45-2022_wirkung-nat-brennstoffemissionshandel.pdf

Umweltbundesamt (UBA). (2023a). Energieverbrauch nach Energieträgern und Sektoren. Retrieved June 16, 2023, from https://www.umweltbundesamt.de/daten/energie/energieverbrauch-nach-energietraegern-sektoren#allgemeine-entwicklung-und-einflussfaktoren

Umweltbundesamt (UBA). (2023b). Erneuerbare Energien in Zahlen. Retrieved June 16, 2023, from https://www.umweltbundesamt.de/themen/klima-energie/erneuerbare-energien/erneuerbare-energien-in-zahlen#strom

Umweltbundesamt (UBA). (2023c). Stromverbrauch. Retrieved June 16, 2023, from https://www.umweltbundesamt.de/daten/energie/stromverbrauch

United Nations (UN). (1992). United Nations Framework Convention on Climate Change: UNFCCC.

United Nations (UN). (2015). Paris Agreement.

United Nations Framework Convention on Climate Change (UNFCCC). (2009). Copenhagen Accord: Decision 2/CP.15. In United Nations Framework Convention on Climate Change (Ed.), *Report of the conference of the parties on its fifteenth session, held in Copenhagen from 7 to 19 December 2009*.

United Nations Framework Convention on Climate Change (UNFCCC). (2022). Nationally determined contributions under the Paris Agreement. Synthesis report by the secretariat: FCCC/PA/CMA/2022/4.

World Resources Institute (WRI), & World Business Council for Sustainable Development (WBCSD). (2004). The Greenhouse Gas Protocol: A corporate accounting and reporting standard. Retrieved June 16, 2023, from https://ghgprotocol.org/sites/default/files/standards/ghg-protocol-revised.pdf

Open Access This chapter is licensed under the terms of the Creative Commons Attribution 4.0 International License (http://creativecommons.org/licenses/by/4.0/), which permits use, sharing, adaptation, distribution and reproduction in any medium or format, as long as you give appropriate credit to the original author(s) and the source, provide a link to the Creative Commons license and indicate if changes were made.

The images or other third party material in this chapter are included in the chapter's Creative Commons license, unless indicated otherwise in a credit line to the material. If material is not included in the chapter's Creative Commons license and your intended use is not permitted by statutory regulation or exceeds the permitted use, you will need to obtain permission directly from the copyright holder.

3

Crafting Corporate Sustainability Strategy

From Integrated Thinking to Integrated Management

Markus Beckmann, Thomas Becker, and Oliver Zipse

3.1 Introduction

The aim of sustainability is to ensure that present and future generations can thrive within the ecological boundaries of our planet (Steffen et al., 2015). As the climate crisis illustrates, however, a linear economy that depletes our natural resources and contributes to global warming threatens to destroy the "safe operating space" (Rockström et al., 2009, p. 472) that allows humanity to thrive. Sustainable development thus requires a change in our current economic model. A different, circular, more equitable, and net-zero future is needed.

Since strategy "is about shaping the future" and "moving from where you are to where you want to be" (Mckeown, 2016, pp. xxi, xviii), a sustainable future requires *Crafting Corporate Sustainability Strategy* in an effective way. On the Road to Net Zero outlined in this book, *strategy* is an important step that connects the previous Chap. 2 and the following Chap. 4 (cf. Fig. 1.1). Chapter 2 introduced the idea of *Science-Based Target Setting* as a means of translating the global challenge of combating climate change to the level of individual company contributions. Science-based targets provide a common

---

M. Beckmann (✉)
FAU Erlangen-Nürnberg, Nuremberg, Germany
e-mail: markus.beckmann@fau.de

T. Becker • O. Zipse
BMW AG, Munich, Germany

language for what it means for a company to be on a Paris-aligned Road to Net Zero. Nevertheless, the targets themselves do not tell a company what that journey looks like. Developing a specific road map and getting all business functions on board to embark on it is what strategy is all about.

The purpose of this chapter is to discuss how companies can craft this type of corporate strategy by systematically integrating sustainability into their strategic analysis, goals, processes, and learning. The chapter is structured in three sections. Section 3.2 starts with a short overview of the strategy concept in management and then considers the drivers that push or pull businesses to consider sustainability before it continues with different options for integrating sustainability. The remainder of that section then discusses the different steps of the strategy process and how sustainability interacts with it. Section 3.3 presents the expert conversation between Prof. Oliver Zipse, CEO of BMW Group, Dr Thomas Becker, VP Sustainability & Mobility at BMW, and Prof. Dr Markus Beckmann, FAU Chair for Corporate Sustainability Management. Section 3.4 then discusses an outlook on the future of integrated sustainability strategies before Section 3.5 concludes with a link to the next Chap. 4 on the *Future of Corporate Disclosure*.

3.2 Strategy Development and Sustainability: Past and Present

In everyday language, strategy describes a plan of action or policy designed to achieve desirable ends with available means. In business, strategy "can be defined as the determination of the basic long-term goals and objectives of an enterprise" (Chandler, 1962, p. 13). Traditionally, strategy research distinguishes between strategy content and the strategy process (Rajagopalan et al., 1993). A prominent approach to structuring the strategy process is to distinguish between its four phases: (1) environmental scanning, (2) strategy formulation, (3) implementation, and (4) evaluation of its performance (Wheelen et al., 2017) (similarly David, 2011, who integrates environmental scanning as a part of strategy formulation). Sustainability requires a systematic integration in each phase. We will discuss each step in more detail below.

Analytical tools, such as Porter's (Porter, 1979, 2008) five forces or Barney's (Barney, 1997) VRIO framework, serve to structure strategy development from a top-down perspective. By contrast, scholars such as Mintzberg and Waters (1985) argue that strategic plans rarely unfold as intended; rather, strategy patterns emerge from the bottom up through individual action and adaptation. While this idea of emergent strategy helps to understand the complexity of strategic learning, for the sake of brevity, this chapter focuses on the planned and deliberate integration of sustainability into strategy.

To survive and thrive in the marketplace over the long term, a firm's strategy is to maximize competitive advantages and minimize competitive disadvantages. Wedding sustainability with a firm's strategy can occur at three hierarchical levels: corporate, business, and functional (Wheelen et al., 2017). Conventionally, a corporate strategy is related to the overall direction of the firm; therefore, it asks where to compete to achieve stability and growth, whereas business strategy focuses on the competitive positioning of products and services in the relevant markets, and functional strategy focuses on leveraging resource productivity by developing distinct competencies in specific functions such as production, marketing, or procurement. At all three levels, sustainability can influence the goals and constraints of a company's strategy.

Various drivers, which can be grouped in different ways, are available for companies to consider sustainability in their strategy (Engert et al., 2016; Meffert & Kirchgeorg, 1998; Oertwig et al., 2017; van Marrewijk & Werre, 2003). These include external/internal drivers, push/pull factors, market/non-market forces, direct/indirect drivers, or supportive/hindering factors (Engert et al., 2016), with some factors falling into several categories simultaneously. One familiar example is customer demand, which acts as an external driver, a market force, and a pull factor (representing an opportunity if realized). Regulation and legal compliance are other external drivers, but they represent non-market forces that act as push factors (posing a risk if ignored). A specific example is the EU's ban on the sale of new petrol and diesel cars from 2035, adopted by the EU Parliament in February 2023 (European Parliament, 2023), which will drastically change the business context for the automotive industry. Other external drivers include investor expectations, transparency requirements, and financial market pressures, which can act as both push and pull factors. Internal factors can include the potential for cost reduction through eco-efficiency gains, top management vision, or employee motivation for sustainability. Supportive internal factors include a responsible organizational culture, professional risk management, competence in quality management that seeks continuous improvement, and a strong capacity for innovation. Hindering internal factors include a lack of resources and competencies, short-termism, and weak leadership. Barriers can also be external, such as poor regulation or a lack of customer demand for sustainable offerings.

Because the specific combination of these sustainability drivers looks different in different contexts and for different businesses, companies have integrated sustainability in a variety of ways that reflect different levels or styles (Baumgartner & Ebner, 2010) or geographic approaches (Burritt et al., 2020).

For simplicity, this chapter distinguishes between three types of strategic sustainability considerations: stand-alone, complementary, and integrated. In a *stand-alone sustainability strategy*, a company addresses social and environmental issues in a way that is not linked to the firm's corporate or business strategy. Here, the company may address sustainability as an intrinsic add-on (for example, by philanthropic projects unrelated to its core business), or it may respond to generic external expectations that are irrelevant to its competitive strategy. In a *complementary sustainability strategy*, sustainability complements the creation of competitive advantage, yet without challenging the existing corporate and business strategy (for example, with eco-efficiency strategies that generate cost benefits but leave the company's product portfolio and overall mission untouched). Finally, *integrated sustainability strategies*, which use sustainability considerations to challenge and potentially redefine a company's corporate strategy (where to compete) and business strategy (how to gain competitive advantage), have the most profound leverage but also the highest level of complexity. In the automotive industry, this could include considerations about changing the powertrain technology portfolio, building secondary material ecosystems, or offering mobility as a service. Finally, within these integrated strategies, companies can integrate sustainability from a more instrumental perspective as "a strategic and profit-driven corporate response to environmental and social issues" (Salzmann et al., 2005, p. 27) or go further and define positive external impact as the purpose of their organization (Bansal & Roth, 2000; Van Zanten & Van Tulder, 2021). Strategy then becomes not only about long-term competitive advantage but also about the "why" and "how" of thriving in the marketplace.

While a list of sustainability drivers and strategy types paints a rather static picture, the idea of sustainability *maturity* or *stages* draws attention to the evolution of a sustainability strategy over time. Maturity models range from the simple distinction of two levels (e.g., laggard vs. leader, Hahn (2013)) to five-stage models (e.g., initial, managed, defined, quantitatively managed, and optimizing, as introduced by Verrier et al. (2016)). Despite these differences, the maturity perspective highlights the external and internal dynamics that influence sustainability strategies. From an external perspective, sustainability maturity reflects the constant evolution of external drivers of sustainability. Regulations change, new technologies emerge, competitive pressures shift, and new customer and investor expectations arise. This is especially true for sustainability drivers. On the one hand, the factual urgency of challenges, such as climate change, biodiversity loss, or resource depletion, is increasing. On the other hand, changes in stakeholder awareness of environmental and social issues drive the political salience and institutional regulation of ecological and social issues. In the automotive industry, for example, increasingly strict regulations on fleet CO_2 emission or human rights due diligence illustrate how this evolution of external requirements demands a corresponding maturity of sustainability integration within the company.

Similarly, an internal perspective on sustainability maturity emphasizes the importance of organizational competencies and their development over time (Dyllick et al., 1997). In the past, in the early stages, many companies responded to sustainability challenges or external criticism with rather limited and often defensive strategies (Meffert & Kirchgeorg, 1998) because they lacked the knowledge and resources to address the issues. However, by investing in early-stage practices, such as environmental compliance, companies gain knowledge, expand their capabilities, and can use them to implement eco-efficiency gains or eventually develop new products and even new markets.

In corporate practice, the idea of sustainability maturity can describe how the focus and scope of sustainability management have changed over the past decades. To illustrate, consider the automotive industry and its evolving focus from cleaner production via cleaner products to sustainable value chains. In 1973, when BMW became the first automotive company ever to appoint an environmental officer, one of BMW's motivations was to respond to the challenge that its manufacturing processes created vibrations that affected the neighboring community. Consequently, the initial focus of sustainability management was local, rather reactive, and focused on the company's own manufacturing operations. Nevertheless, establishing systematic environmental management led to significant improvements in *cleaner production* and created valuable eco-efficiency capabilities. In the ensuing decades, BMW has consistently continued to reduce its manufacturing emissions and improve resource efficiency, and it now bases its sustainability strategy on the "LEAN. GREEN.DIGITAL." principle for all of its plants. To reap the sustainability benefits of these competencies, BMW plans to have its Debrecen, Hungary, plant operational by 2025 as the company's first carbon-neutral factory.

While cleaner production initially focused on local emissions and employee safety in a company's own operations, the *cleaner product* perspective has since shifted the focus to the environmental performance of a product during its use. For the automotive industry, customer expectations and regulatory requirements have demanded significant improvements in fuel efficiency and on-road emissions. This includes both CO_2 emissions, which contribute to global warming, and pollutants, such as particulate matter or nitrogen oxides, that affect local communities. In response, companies have invested in cleaner and more efficient drivetrain technologies, including improvements to internal combustion engines and the development of new powertrain technologies, such as plug-in-hybrids, battery electric vehicles, and hydrogen-powered cars. These new products and product portfolios reflect how deeply sustainability considerations are now being integrated into business strategy. To remain competitive, lead in terms of sustainability, and meet future regulatory demands, companies are formulating strategic targets for their own products. In the case of BMW, the company committed to reducing CO_2 emissions per car and kilometer driven by at least half of the 2019 levels by 2030 (BMW, 2021).

In addition to taking responsibility for a company's production and products, mature sustainability strategies today also manage the company's responsibility for its value chain. Creating a *sustainable value chain* further extends the scope of the sustainability strategy from internal processes to the entire life cycle. This includes environmental and social issues, including human rights, both upstream (such as labor and environmental questions in the extraction of raw materials) and downstream (disposal and recycling). Companies are embracing value chain responsibility (Baier et al., 2020) for the ethical sourcing of critical resources, and they are responding to external drivers, such as customer expectations and increasing regulation (e.g., the German or EU supply chain due diligence regulation).

In the automotive industry, a strategic approach to sustainable value chains is also needed to meet the ambitions of a net zero future. To date, emissions targets have mostly focused on tailpipe emissions; that is, the direct CO_2 emissions of a car on the road. The transition to electric or hydrogen mobility can eliminate these emissions during the use phase, but it shifts the focus to emissions at other stages of the life cycle. These include the energy and emissions of battery production, the sourcing conditions (including human rights impacts) for critical battery and drivetrain materials, such as lithium, cobalt, nickel, and rare earth elements (Schmid, 2020), the sourcing of electricity for car usage, and the recycling of batteries.

A value chain-oriented sustainability strategy goes hand in hand with the idea of circularity (Ellen MacArthur Foundation, 2013). Closing the loop (for example, through the use of secondary materials) is critical to reducing emissions and securing the availability of scarce resources. While a value chain-oriented approach can increase the sustainability impact and business benefits, it also increases complexity. This type of holistic sustainability strategy must involve all related corporate functions (e.g., production, R&D, procurement, logistics, and marketing), collaborate with partners along and across the value chains (e.g., suppliers, data providers, and auditors), and allow partnering with non-market stakeholders (e.g., the charging infrastructure for electric mobility) (Beckmann & Schaltegger, 2021). Against this background, sustainability has implications for virtually all aspects of management, thereby requiring a much more integrated approach to strategy. Indeed, sustainability requires a systematic integration of all steps of the strategy process: (1) environmental scanning, (2) strategy formulation, (3) implementation, and (4) evaluation of its performance (David, 2011; Wheelen et al., 2017).

In the first step, *environmental scanning* gathers information about the relevant external environment (such as natural resources, regulation, and industry analysis) and internal environment (such as the organization's current capabilities). The case of climate change illustrates the importance of systematically including sustainability aspects at this stage. For companies, climate change poses a variety of risks, ranging from regulatory risks (e.g., bans on internal combustion engines) and supply chain risks (e.g., water scarcity in raw material production) to physical risks (e.g., the impacts of extreme temperatures on the operability of battery electric vehicles). Therefore, a thorough and, where possible, scientifically based understanding of the climate system is key to subsequent strategy development. An example of an increasingly critical environmental parameter is the remaining carbon budget, which humanity must not exceed to limit global warming, as agreed upon in the Paris Agreement. For many companies, non-market actors, such as the Intergovernmental Panel on Climate Change (IPCC), are now becoming relevant stakeholders.

Since climate change is not the only sustainability challenge, environmental scanning is needed to capture the full range of social and ecological issues of strategic relevance. Moreover, companies cannot address all issues simultaneously and with equal emphasis. In fact, any strategy requires the prioritization of what matters most. Materiality analysis is a relevant tool for this type of prioritization (Whitehead, 2017). In the field of sustainability, materiality analysis is often based on the combination of a company's internal perspective (what matters to the company) and the external assessment of its stakeholders (what matters to the world). While win–win issues (such as eco-efficiency) may have direct financial materiality, "tensioned topics" that (still) lack a business case but have a societal impact (Garst et al., 2022) may have strategic business relevance in the medium and long terms. This idea of "dynamic materiality" (Kuh et al., 2020) highlights that the environmental scanning phase requires a more systematic interaction with diverse stakeholders, including both market stakeholders, such as customers, investors, and suppliers, and non-market stakeholders, such as scientists, NGOs, and regulators.

The second phase of the strategy process, *strategy formulation*, consists of several steps. First, a company clarifies its mission (Wheelen et al., 2017) to consider where sustainability considerations can significantly shape its understanding of why it exists and operates in the market. For the sake of brevity, however, we focus on the two steps of formulating *strategic objectives* and the *strategic plans* needed to achieve them.

In the context of sustainability, current integrated approaches to strategic objectives increasingly use the formulation of *Science-Based Targets* (SBTs). In the case of the climate debate, SBTs offer an emerging approach to align corporate emissions with the temperature target of the Paris Agreement (Bjørn et al., 2022). SBTs are gaining in importance for several reasons. For the overall goal of combating climate change, the appropriate allocation of the remaining carbon budget to individual sectors and companies is important. Appropriately identified SBTs could thus help promote global emission reductions. For companies, however, having reliable targets that allow them to plan and that are respected by external stakeholders is important. The more robust SBTs and their underlying methodology, the better companies can use them to quantify sustainability goals and track their implementation. While SBTs for climate change have received the most attention to date, the basic idea is also relevant to other sustainability issues, such as biodiversity. In any case, the formulation of specific SBTs requires intensive stakeholder engagement to translate global system goals to the corporate level (Andersen et al., 2021).

As a critical next step in strategy formulation, companies develop *strategic plans* (Wheelen et al., 2017) that outline how the mission and strategic objectives will be achieved. For an integrated sustainability strategy, this step is characterized by additional complexity due to the assumption of responsibility for the entire value chain. In the case of a climate strategy that formulates SBTs, companies need to consider emissions along the entire value chain. This requires disaggregating total emissions into Scope 1 emissions (arising from the company's own operations), Scope 2 emissions (arising during the production of energy procured by the company), and Scope 3 emissions (arising in the value chain) (Kaplan & Ramanna, 2021). For strategic planning, a significant difference exists in terms of the actions taken to reduce these emissions. For Scope 1, companies need to understand and change their own operations; for Scope 2, they can change their energy procurement; and for Scope 3, they need to engage with their suppliers and incentivize or actively help them to decarbonize their processes. To illustrate, BMW has already contractually agreed with more than 400 suppliers to use 100% green electricity by 2022. Similarly, pilot projects are pioneering the production of CO_2-reduced steel, as this production replaces coal with natural gas, hydrogen, or green electricity (BMW, 2022). Strategic planning for sustainability therefore requires a much deeper interaction with suppliers and other stakeholders.

Stakeholder engagement can be used to identify the biggest levers for CO_2 reductions and to analyze the feasibility of measures outside a company's organizational boundaries.

In the third phase, *strategy implementation*, strategic plans are put into action. In traditional business strategy, this phase involves implementing programs and tactics, allocating budgets, and carrying out the procedures to get the job done (Wheelen et al., 2017). While this is also true for sustainability, an integrated sustainability strategy adds complexity and requires an even more integrated management approach. Because sustainability has multiple dimensions that interact and cannot be managed in silos, it requires the alignment of different departments and the organization of cross-functional collaboration (Baier et al., 2020). To do this, companies need adequate data and information. An integrated management approach to sustainability therefore relies on appropriate indicators that are measured, shared, analyzed, and made available throughout the organization, and even to value chain partners. In addition, an integrated approach to management allocates resources and incentives in a way that is aligned with long-term sustainability goals. To ensure that improvements in one sustainability dimension are not incurred at the expense of other sustainability or business objectives, integrated management is needed to identify potential trade-offs and to provide guidance on how to address them (Baumgartner & Ebner, 2010). Since sustainability measures are often investments in future benefits, an integrated management approach is also needed to align individual budgets and incentives with these long-term goals. Measurable sustainability indicators then become performance criteria for management compensation.

The final phase of the conventional strategy process is the *evaluation and control* phase, which monitors performance. At the same time, the evaluation phase does not end the strategy process; rather, it provides feedback for an iterative engagement with all previous phases (Wheelen et al., 2017). This feedback and control is an important internal function for the company. In the case of an integrated sustainability strategy, the evaluation phase also generates information for reporting a company's sustainability performance to an external audience. Over the past few decades, sustainability reporting has evolved from a voluntary practice to a de facto standard and subsequently to a regulatory requirement for most multinational companies. However, until recently, most companies reported their ESG indicators in separate reports, which did not give the indicators the same prominence and assurance as financial data. However, in an integrated sustainability strategy that aligns different stakeholders and sustainability dimensions with business strategy, aligning these different perspectives by marrying sustainability and financial reporting becomes important. This is what integrated reporting is all about (Churet & Eccles, 2014). For investors, integrated reporting is about providing the transparency needed to make sustainable investment decisions. For companies, its aim is to overcome internal silos and strengthen an integrated approach to strategy and management (Higgins et al., 2019). So far, however, integrated reporting has not yet become the new reporting norm. When BMW combined its Annual Report and Sustainable Value Report for the first time in an integrated BMW Group Report in 2021, it became the first premium automotive company worldwide to do so. Given the dynamic developments in sustainability reporting standards and regulations, it still remains to be seen which specific frameworks and reporting approaches will evolve. Crafting corporate sustainability strategy for the future will therefore interact with the future of corporate disclosure (see Chap. 4).

3.3 Expert Conversation on Integrating Sustainability into Corporate Strategy

What Are the Drivers for Integrating Sustainability into Corporate Strategy?

Beckmann: When I took a tour of the BMW plant 10 years ago, I learned that BMW was the first automaker to appoint an environment officer in 1973. At the time, one of the drivers for the new role was that the heavy machinery at the plant was causing vibrations that were a concern to the plant's neighbors. The Environment Officer took up these local community issues and helped translate them into improvements at the plant. In 1973, BMW's sustainability management began with one specific driver. Fast forward to today, and BMW has a much more sophisticated sustainability management system in place and is striving to be the most sustainable premium provider of individual mobility. What are the key factors driving you toward this goal today?

Zipse: Sustainability is a moving target. When we introduced the Environment Officer some 40 years ago, it was a separate role that took care of sustainability alongside the core business. This has changed radically, and I would identify four main drivers of sustainability.

Beckmann: What are these four drivers?

Zipse: First, society is changing. Environmental issues are constantly changing, and society has a different awareness of them today than it did 40 years ago. The second point is policy and regulations. Regulatory policies are changing and getting much harder for the industry. The third point is that our financial system is much more targeted toward ecological and sustainable performance. That's a new aspect and a quickly accelerating one.

Beckmann: What is the fourth key driver of sustainability?

Zipse: Last but not least, the fourth sustainability driver is changing customer behavior. Our customers' desire to buy a product, to spend money, is very much linked to a sustainable image and a sustainable product substance. Therefore, all four of these drivers make us rethink, or forward think, how we develop our corporate strategy. As the essence of all these four points, we are putting sustainability right into the core of our corporate strategy.

Beckmann: Can you give an example?

Zipse: Sure. Take our transition to integrated reporting. As of 2021, we no longer issue separate reports. Therefore, we no longer have one report for the business and financial community and another report for NGOs and society. There is only one report. Having an integrated report is also a disciplinary tool. Whatever we do and communicate must be verifiable, measurable, and true. In the automotive industry, we are the first company to combine the sustainability report with our regular BMW Group report into a single report. This is a significant step for us—and it also shows that sustainability is not a fixed target but constantly moving.

An Integrative Approach: How Does It Affect Management?

Beckmann: Integrated reporting addresses the diverse stakeholders you have: not only your financial investors but also your regulators, customers, NGOs, etc. Integrated reporting is about addressing these different external views together. When integrated reporting was first introduced, another idea behind it was to break down the silos that companies have internally—sustainability department, finance department, reporting department, and so forth. Do you see this internal integration reflected in the way you approach your sustainability strategy?

Zipse: The integrated report reflects what we do internally. The best solution is that every stakeholder—our employees, our shareholders, and our management—is integrated into our decision-making processes. The time when sustainability was seen as an extra is over. Instead, we are and must be intrinsically motivated to build sustainability right into the product.

Beckmann: Why is that relevant?

Zipse: Every decision we make today will affect the market for the next 12 years. The products we configure now, in 2021, will not come to market until 2025 at the earliest. However, the same architecture usually lasts for two consecutive products, which brings us to 2040. So, whatever we do today has to be market ready for the entire product life cycle. That is why leapfrogging into the future is so important now.

Beckmann: You take a long-term perspective because the development time and the life cycle make it important to put sustainability considerations at the beginning of the development process, not at the end. Can you give concrete examples of how developing this long-term approach differs now compared to the early 2000s, especially concerning this integrative management approach?

Becker: The critical issue we face in the coming years is to look beyond production, which was the beginning of sustainability as a function. We also need to look beyond the product, which was the issue around emissions and electrification. Instead, we need to look deeper into the value chain. This creates a twofold challenge. One is to integrate the right objectives into product planning right from the outset, using the right mechanisms. We are already doing this intensively for our future products by asking: How can we reduce the CO_2 footprint of inputs such as steel, aluminum, or high-voltage storage devices? To achieve this goal, the first step is to negotiate with our suppliers to source the right energy.

Beckmann: What is the second step?

Becker: The next one will be to source much higher percentages of secondary materials, which will give us an additional option to reduce our footprint, not only in terms of CO_2 but also in terms of the resources needed.

Beckmann: So the principles of circularity are important.

Becker: Absolutely. All of this is good, and we need to do it. But to credibly demonstrate what we have achieved, we now need to build up the reporting, target setting, and steering mechanisms so that we can subject our environmental footprint numbers to the same level of scrutiny as our financial numbers with our certified account. This is why integrated reporting is so essential. An integrated approach to sustainability is a massive challenge, because it goes far beyond our own organization. It extends into our supply chain. This is something we are actively tackling at the moment.

What Value Does Sustainability Deliver as an Overarching Corporate Strategy?

Zipse: From an academic perspective, where do you see the value for companies in integrating sustainability into their strategy?

Beckmann: There is a short and simple answer and a long and complex answer to this question.

Zipse: What is the short and simple answer?

Beckmann: The quick response is that you have the business benefits of managing the risks, costs, revenues, and license to operate. As you just described, companies need to respond to changing regulations to maintain their license to operate in the marketplace. Regarding risk management, ignoring sustainability can lead to litigation risks, operational risks, reputational risks, and so on. Then, you can manage costs. When you reduce waste, you conserve sources. After all, waste is, by definition, wasteful. In manufacturing, material and energy efficiency can save money and go hand in hand with lean management. Finally, you can be more attractive to the stakeholders you want to work with, such as employees, investors, and, of course, consumers. Products with greater sustainability can help attract consumers, drive innovation, and create new market opportunities.

Zipse: How is the long answer different?

Beckmann: These sustainability drivers are the classics, but they are also quite generic. When it comes to concrete strategy, the answer is more complex. What matters here are the specific company and the context-specific causal pathways that can translate a particular sustainability issue into one of your performance drivers. The impact logic may vary depending on the respective industry, the position within the industry, or the maturity of the sustainability strategy.

Zipse: Can you illustrate what that means?

Beckmann: Take the example of material and energy efficiency in manufacturing. BMW's Green.Lean.Digital production has come a long way in this regard. By contrast, for other companies that are sustainability beginners, this is still low-hanging fruit with a relatively short causal chain. If you have an energy-intensive production, you can do an eco-efficiency analysis, implement more efficient solutions, such as heat recovery, and save emissions and costs. This yields quite straightforward, simple, and measurable results in the short term.

Zipse: I see. What would be an example of a more complex situation?

Beckmann: When developing a strategy for future scenarios with high levels of complexity and uncertainty, the pathways that link sustainability to business success are much more intricate. For example, when planning the Road to Net Zero, achieving carbon neutrality with business benefits is anything but straightforward, simple, or easy to implement with certain short-term results. On the sustainability side, effective decarbonization is complex. Where do your emissions occur over the life cycle, and where are the best places to reduce CO_2? How can you collaborate with others? On the business side, how do you translate those CO_2 savings into business benefits? How can you compare or even put a price tag on different options, given the uncertainties of future regulation, charging infrastructure, or market demand?

Zipse: These are indeed more complex questions.

Beckmann: And they interact. Because multiple factors, such as future government, customer, employee, or investment behaviors, play a relevant role, different sustainability drivers interact in multiple ways. Strategy then becomes an issue of understanding, selecting, and creating favorable causal relationships between sustainability and business success and raises questions such as the following: Given your current position, what configurations allow you to align sustainability and business success? Can you influence external parameters, such as future regulations or industry standards, or do you have to take them for granted? Given your competencies, can you innovate new technological solutions or create market demand? The answers to these questions will differ for different companies at different stages of their sustainability maturity.

Zipse: Do you know of any good examples—perhaps even outside the automotive industry—where this integration into corporate strategy and philosophy has been successful?

Beckmann: There are many inspiring examples in different contexts. Let's take a look, for example, at a sustainability pioneer in the textile industry: Patagonia. They have always been sustainable. They have a sustainable customer base. They have a sustainable story. But they operate in a niche. Therefore, it is challenging to use Patagonia as a role model for a company with a mass market position or a broad customer base.

Zipse: If sustainability has always been at the core of a company's business model, the transformation path will likely be faster. What are some examples of companies in other industries that have undergone a more fundamental change?

Beckmann: When incumbents in traditional markets transform their business models, some react to disruptive change, such as in the case of scandals. Others anticipate change proactively. For instance, in the food industry, I like the example of "Rügenwalder Mühle," a family-owned company. Without a crisis forcing them to do so, they are currently disrupting their meat-only business and are developing plant-based products as a second, alternative business model.

Zipse: I know this example. It is indeed interesting.

Beckmann: I agree. With more and more people wanting to go vegetarian or even vegan, this strategy makes the company fit for the future. It is also helping to transform the market. Its significant growth in plant-based products gives consumers more choices of tasty and sustainable products. At the same time, the company continues to offer meat-based products, but aims at higher standards by focusing on more animal welfare-oriented production. In this way, the company is developing valuable options for success in the food market of the future. In a way, this is what sustainability is all about: acting in ways that increase our options for the future. For companies, this may mean disrupting their current business models.

How Can We Close the Gap Between Intentions and Behavior When It Comes to Sustainability?

Beckmann: One of the conceptual challenges within sustainability is that we often say—and rightly so—that sustainability takes a comprehensive, holistic perspective. But if everything matters, then strategy lacks focus. How do you prioritize your sustainability goals? How do your priorities change over time?

Zipse: Whatever you do, you must have a comprehensive or 360-degree approach to strategy. It is easy to pick a specific product, a specific market, or a specific drivetrain and run a prototyping exercise on it. I think we are beyond that. We launched the BMW i3, for example, in 2013. Now, almost 10 years later, it is all about learning how the world—the customers—respond to the product and then very quickly integrating that into your strategy. There is a big difference between what people say and what they do—a big difference between people's statements about sustainability and their actual buying behavior.

Beckmann: In academia, we call this the intention–behavior gap. I always try to teach this to my students.

Zipse: I assume that this academic description is somewhat consistent with our observations in real life. At the end of the day, a company like BMW must also be financially successful. This is not just a sustainability issue. It is about understanding the buying behavior of your customers. We now see that sustainability is becoming what you call the "license to operate." People will not buy individual mobility that is not sustainable. And that is changing rapidly. But, at the same time, the market does not change overnight. There is a very long transformation period.

Beckmann: That sounds like a balancing act.

Zipse: Indeed. The tricky part is to balance the fact that every year, every month, and every week that there is both conventional and progressive customer behavior. We serve young people and older people. There are digital natives and people who do not care about digital functions in the car. So, the trick is to understand, acknowledge, and serve these different needs simultaneously. Our answer is that your product development, your production strategy, and your marketing strategy have to be flexible. We serve 196 markets, and it will come as no surprise that each market behaves differently. Even within one market—say the European market—we find huge differences in buying or customer behavior. Customers in Oslo typically behave differently than those in Sicily. However, we serve them as the same company. Flexibility in all your processes and the ability to react quickly to market changes are key.

To What Extent Do Different Stakeholder Needs Change Sustainability Goals?

Beckmann: Customer behavior is constantly feeding back into your strategy. Your customers are different from other OEM [Original Equipment Manufacturer; here: car makers] customers. Do their expectations—and, more importantly, the specific competencies that distinguish BMW—influence your sustainability strategy? How does your sustainability strategy differ from that of your competitors?

Becker: We take a close look at the differences between the markets. The notion of sustainability in Beijing is very different from that in Copenhagen or Los Angeles. While there are things that we obviously need to do across the entire spectrum of our products and that need to be deeply rooted in the entire organization and the processes, the actual customer expectation doesn't necessarily have to be the same. For example, we need to be able to give to every customer, wherever they are, accurate information about the footprint of our product. However, the way this is valued or demanded differs. While sustainability in Europe is very much about demonstrating that a product is safe in every respect, the Chinese perception is much more about personal experience, entertainment, and the direct benefit to the customer. We have to take that into account.

Beckmann: So, it's a lot about what you do, how you communicate it, and how you integrate the voices of different stakeholders?

Zipse: Transforming into a truly sustainable company starts with the right collaboration. We know that an automotive company cannot do it alone. You depend on the charging infrastructure. You depend on city operators. You need all your suppliers. You depend on digital companies. What is your experience there? What are the big stumbling blocks in this "need to collaborate?"

Beckmann: When we look at collaboration for sustainability, one stakeholder group that I am particularly interested in is a company's competitors. Many sustainability issues are systemic in nature. They are not specific to one company but relate to the industry as a whole. From a strategic perspective, many of these issues have a pre-competitive character. They are relevant to the industry as a whole but are not necessarily a source of individual competitive advantage. Few customers understand or care about the details when it comes to technical issues, such as the banning of certain hazardous substances or adoption of specific technology standards. However, customers do care when a major scandal occurs in the industry. In extreme cases, the entire industry gets a bad name, such as in the Dieselgate scandal. There is a need to work together on industry-wide solutions, such as shared standards.

Zipse: Let me make a quick comment. We don't talk about Dieselgate. It was a 'gate' involving a particular company. It was not a 'gate' of the technology.

Beckmann: I am not questioning that from your internal perspective. However, the impact on the public perception of the technology was severe. What happened to one company affected other companies.

Zipse: I just wanted to reiterate that we were not part of the primary root cause. But, yes, it affected the entire industry, and it affected us, too. But let me go back to your initial point about collaboration. What are your thoughts on that?

Beckmann: When you look at sustainability issues in the value chain where competitors are sourcing from the same or similar suppliers, common standards can help everyone increase transparency, reduce complexity, and lower transaction costs. However, when competitors cooperate, one of the challenges is to respect antitrust regulations. The idea is not to restrict competition but to create a level playing field. Once the rules of the game include appropriate sustainability standards, companies can compete on how best to innovate from there. Therefore, considerable potential exists for competitors to work together and to include suppliers, NGOs, and intermediaries to drive the sustainability transformation of the entire industry and its ecosystem. Ideally, this collaboration across the ecosystem contributes to sustainability and makes the industry fit for the future.

Would an Ecosystem Approach Be a Strategy for Rapid Technological Development?

Beckmann: You talked about infrastructure for charging, city management, parking, and traffic management. If you try to create solutions here, you are operating and innovating in an ecosystem. What role does cooperation play here?

Zipse: In the current technological development picture, the cost factor is becoming more and more critical. Today, you can build almost everything into a car: autonomous driving, driverless vehicles, etc. These features are no longer a question of technical feasibility, but of commercial viability. What kind of technology do you put into the car so that you still have a product that is viable in the market and generates a positive margin?

Beckmann: How does this focus on creating viable products relate to the ecosystem and partnership perspective?

Zipse: The reason is that scaling becomes critical to the cost of the vehicle. When you get into new technologies, you are well advised to find a partner who can help you scale—not only in your own product range, but also to reduce costs across different car manufacturers. Then, you have the need for a battery cell. You have the same issue with cameras for automotive autonomous driving functions in the car or anything that has to do with connectivity in the car. Then, of course, you have to find the right partners. What you find out now is that these are often not the traditional OEM suppliers—the so-called first tiers. Instead, they are new entrants to the industry.

Beckmann: For example?

Zipse: Take the battery cell. Our major battery cell manufacturer, CATL, did not even exist 10 years ago. In this dynamically changing ecosystem, partnerships are essential to get the right technologies and have them at the right cost base.

Becker: If you look at the sustainability leaders in other industries and at how they position their product, almost none of them are going to tell an "I'm so fantastic, I did it all" story. In many cases, credibility also comes from working with others, pooling competencies, organizing value chains properly, and engaging your suppliers. Keeping this in mind can be highly relevant to sustainability success.

How Can Corporate Sustainability Goals Contribute to Society?

Beckmann: You just talked about the importance of working across the entire value chain with your new technologies. You need to work with your value chain partners to manage the cost, complexity, and life-cycle impacts of new technologies. But when we discussed sustainability drivers earlier, the first one you mentioned was changing societal attitudes. Stakeholders representing this shift are NGOs, social movements, and think tanks. Do you see a difference in the way you work with those partners—not just talking to or listening to them, but incorporating their ideas, opening up the innovation process, and piloting solutions—compared to the way you work with the traditional customer or first, second, or third tier suppliers in the value chain?

Zipse: For many years, we have gone far beyond defending our business model against those who do not understand or support it. We listen to all the stakeholders around us. We talk to customers, of course. We talk to our suppliers, we talk to parliamentarians, we talk to NGOs, we talk to politicians, and we take all that knowledge and put it into our strategy. At the end of the day, we will not meet everyone's taste. However, we increase the likelihood that our products will fit into society. That is important. Of course, what the outside world cannot see is that we have to build this on a functioning business model where you have a contribution margin, price tags on your products, and production costs. But we would not take the easy way out and use that as the only argument for bringing a product to the world. It is also about our contribution to society. In the broadest sense, creating value is a core mission of any large company, not just making a profit.

Beckmann: I fully agree. The primary contribution of companies to society is to create value. They do that by addressing societal needs and improving our ability to meet those needs. To do that, you want to be responsive, you want to see what kind of value is needed, and you want to align your operations with that. However, very often, you cannot do it alone because you need resources and the participation of others. How do you get that input from different stakeholders?

Becker: Coming back to the value chain: You could potentially reduce the CO_2 footprint of a ton of aluminum by 80% if you use secondary materials. It seems obvious to do this as soon as possible. The problems, unfortunately, are the technical performance and quality requirements. For automotive applications, the copper contamination in aluminum must be less than 2%. Can you find suppliers of scrap aluminum that meet this requirement? Is the sorting technology powerful enough to remove the copper? Not yet. However, as soon as you say, "We want more of the high-quality stuff," the question will come back to you because a critical barrier to adequate recycling is the way metal components are currently built into cars.

Beckmann: Can you give us a specific example?

Becker: Take the wiring harness, which is mostly copper. How do we need to install it so that it can easily be removed before the vehicle is scrapped and gets shredded into tiny particle size? To move this agenda forward, you need to find the right solutions with different value chain partners and across industries. Finally, there is an important systemic factor. All decarbonization efforts would benefit greatly if CO_2 prices were predictable and would reliably change the price ratio between primary and secondary materials. As you can see, these things are very much intertwined. We have to understand this and accept that not everything is certain and predictable today. Our task is to maintain our ability to steer economically and efficiently as we move forward.

Zipse: Our products are a collection of 16,000 parts from more than 4000 suppliers—and we are responsible for them. The Supply Chain Act has put that into legal terms. Consequently, if a problem emerges, everybody has the right to say, "You are the aggregator of this car. I'm holding you responsible for the supply chain behind it." Suddenly, the aggregator, as the seller of that car, has to figure out how to organize responsibility across the entire supply chain.

Does a Common Language for the Entire Industry Help?

Beckmann: That makes sense. To aggregate data and orchestrate change across complex value chains, you need a common language to communicate sustainability requirements, measure performance, share data, and drive improvements. For example, having the right metrics in place could give you and others more mileage and more leverage for sustainability. How do you develop this type of common language at the value chain or industry level? How do you organize accountability for the social and behavioral frameworks with your suppliers?

Zipse: You need collaboration to organize upstream and downstream value chain responsibility. You cannot do it alone. Putting it in a contract is not enough. You have to create some kind of transparency. We have brought together 20 German companies—around SAP, Bosch, and other OEMs like Mercedes—to form an automotive alliance that is building a digital transparency chain across many companies. This makes it easier to document, for example, the carbon footprint or quality issues in your supply chain. Involving partners in your business model is crucial.

Becker: You also need to ensure the acceptance and credibility of your products and processes, which we discuss in detail in our annual report. For instance, the aggregation of CO_2 over forty steps in our value chain must produce a correct result. Again, we need to work with others.

Zipse: An essential aspect of being "responsible"—and what makes it so difficult at times—is staying profitable. Profitability is part of our responsibility. Suppose you are running a company in a financially irresponsible way. In that case, you are making a big mistake: You are taking away all your freedom to actively develop the company and actively manage other factors. In order to secure this freedom to act, the company must remain profitable at all times. You cannot take a break for 3 or 4 years. Only with profit responsibility do you have the strength to put resources into innovation and the next step of sustainability.

Is Integrated Reporting the Key to a Unified Strategy?

Beckmann: Ideally, investing resources in innovation and sustainability is also a good investment in future profits. However, when you look at ambitious sustainability goals like a net-zero future, the implementation is a marathon, not a sprint. First, you have to invest in new technologies, knowledge, management systems, relationships, infrastructure, and so on. And you have to take a long-term view. A common criticism of publicly traded companies that respond to stock market expectations is that they focus only on the next quarter's results. How do you reconcile that with a long-term sustainability strategy?

Zipse: What is happening now is that different stakeholder interests are merging. We no longer have a financial stakeholder view that is isolated from an NGO view. These views are merging. Large investment companies are bound to invest in sustainable companies. To do that, they need proof that we are sustainable. NGOs are demanding the same thing. When we still had separate reporting, addressing these different stakeholders with different data was a barrier to aligning stakeholder interests. Therefore, we have now moved to integrated reporting.

Beckmann: What is needed to make this integrated view successful?

Zipse: The trick to aligning stakeholder expectations with a long-term strategy is always to put customers first. They are the lever that keeps your business profitable and provides the foundation for everything else. Your strategy discussions must never lose sight of the customer. For this reason, product development and marketing efforts are essential for understanding your customers in every part of the world, not just through your domestic lens. We see a trend toward diverging product demands: A Chinese customer wants something different—much more digital—than the average German or Central European customer would request. And the American customer has different expectations yet again. This is a critical point. Key sustainability expectations are converging. Customer needs are diverging. Bringing both aspects together in a single corporate approach is the art of strategy.

Beckmann: Sounds like a fascinating journey. Thank you for bringing all these different threads together and giving us the opportunity to discuss them.

3.4 The Future of Integrated Strategies: Challenges, Opportunities, and Key Questions

Integrating sustainability into strategy has important implications for all steps of the strategy process. Sustainability raises additional questions for a company's situational assessment, strategy formulation, strategy implementation, and strategy evaluation and control. This integration goes hand in hand with new opportunities, challenges, and future questions that arise at the intersection of sustainability and other megatrends.

Challenges of Integrated Sustainability Strategies
Sustainability highlights additional social and environmental realities, their systemic interdependencies, and the role of the diverse—and often conflicting and changing—stakeholder expectations related to them. Against this background, the integration of sustainability into strategy can be discussed in light of the challenges of strategizing in a world characterized by the features of volatility, uncertainty, complexity, and ambiguity (VUCA) (Bennett & Lemoine, 2014).

Volatility mirrors the fact that sustainability is a moving target. Social issues, such as human rights concerns in the deep value chain or the massively burgeoning issue of biodiversity conservation, are emerging as material issues that were not as apparent on the radar screen a few years ago. In the area of sustainability, volatility is driven by both rapid changes in the physical environment (as the effects of climate change and ecosystem degradation reach local and global tipping points) and disruptions in the social environment (as customer expectations shift, regulations change rapidly, or new environmental activist groups emerge). In recent years, the pace of change has accelerated, not slowed, thereby increasing the volatility of sustainability issues.

Uncertainty refers to how easily (or not) we can predict the future. Sustainability increases the difficulty of predicting the future with confidence because of its multiple systemic interdependencies, which often behave in nonlinear and surprising ways, including displaying irreversibility. A highly relevant example is the current and future changes in our climate system. Many companies have committed to a climate strategy in line with the Paris Agreement by pledging to reduce emissions in line with the 2 or 1.5 °C target. As discussed above, a fully integrated sustainability strategy benefits from SBTs that translate the remaining global carbon budget to the company level. However, as global warming brings us closer to critical tipping points (such as the thawing of permafrost or the dieback of the Amazon rainforest), climate dynamics may change significantly. In fact, each IPCC report updates the remaining carbon budget by incorporating the latest physical science and other aspects, such as economic growth and the degree of decarbonization achieved. This multifaceted uncertainty creates difficulties for companies today in setting a robust SBT that allows for long-term planning while recognizing the uncertainty associated with the climate and its future evolution. Given the difficulty of accurately predicting long-term systemic interdependencies, sustainability therefore adds to the uncertainty that strategy must address.

Complexity reflects the number of factors that strategy must consider, their breadth and diversity, and their interactions. As complexity increases, comprehensive analysis of the environment and understanding the big picture become more difficult. In the context of sustainability, one reason for complexity is the multidimensional nature of sustainability. To illustrate, consider the United Nations Sustainable Development Goals (SDGs). The SDGs include 17 goals and 169 more specific targets. To measure the achievement of these targets, the UN has defined 231 unique indicators (United Nations Statistics Division, 2022). Complexity arises from the challenge of generating, collecting, and sharing these comprehensive types of data—and, more importantly, analyzing how the different factors relate and interact. By highlighting additional factors, sustainability increases the complexity of strategy development.

Ambiguity can be defined as a lack of clarity about how to interpret a situation. Ambiguity arises when competing interpretations are possible. It occurs when information is incomplete, fuzzy, or contradictory. In the context of sustainability, ambiguity often emerges when companies deal with different stakeholders who have different interpretations of the same issue and whose expectations go in opposite directions. Ambiguity also arises in the aforementioned intention–behavior gap, where customers demand sustainable products but do not actually purchase them. An integrated strategy must make sense of this type of conflicting information. More importantly, it must reconcile conflicting stakeholder views in a way that overcomes perceived trade-offs through innovation (Beckmann et al., 2014). Because the multi-stakeholder orientation and multi-dimensionality of sustainability increase the likelihood of incomplete and conflicting information, sustainability can add ambiguity to an integrated strategy.

Opportunities for Integrated Sustainability Strategies

While the discussion of the VUCA world often focuses on its challenges, the idea of strategy is to play an active role in shaping a company's future context in a way that unlocks new opportunities. Sustainability can create opportunities across all VUCA dimensions. Volatility means that rapidly changing stakeholder expectations and emerging sustainability issues create new search fields for innovation. Moreover, management research has long embraced the notion that uncertainty creates opportunities for leadership and entrepreneurship as both represent practices of uncertainty absorption (Bylund & McCaffrey, 2017; Waldman et al., 2001). According to this logic, sustainability leadership and sustainable intra- and entrepreneurship can provide companies with a competitive advantage in navigating the VUCA world. Companies with authentic and credible sustainable purposes will have an easier time mobilizing this potential. Complexity emphasizes that companies can combine a broader set of factors in their innovation process, allowing the companies to rethink inputs, processes, and outputs in new ways. Finally, the ambiguity that arises from conflicting stakeholder views and incomplete information can represent an opportunity to build novel business models and stakeholder networks that actively align previously competing interests.

Based on a proactive response to the VUCA world, an integrated sustainability strategy can deliver the multiple business benefits discussed at the beginning of this chapter. An integrated sustainability strategy can lead to technology and process optimizations that result in cost savings, improved performance, and increased resilience. By responding to future customer needs, sustainability can add a price premium to a product, increase customer loyalty, and open up new markets. Similarly, an integrated sustainability strategy can serve to improve employee appeal, attract sustainable financial investments, and increase supply chain resilience. At the corporate level, sustainability can secure a company's license to operate and strengthen its competitiveness. At the industry level, driving more sustainable value creation secures the license to operate across the entire ecosystem. To achieve these benefits, the integration of sustainability into strategy must be based on intensive learning, innovation, and change management. An added benefit of a successful integrated strategy is therefore the improvement of organizational agility and adaptability.

Future Questions for the Alignment of Integrated Sustainability Strategies

Integrated strategies focus on the long-term alignment of sustainability and business objectives. This type of alignment raises several follow-up questions related to both sustainability-specific aspects and other megatrends in business.

How Can Sustainability Strategy Be Aligned with Different Time Horizons?

Sustainability requires a long-term perspective. An integrated sustainability strategy requires aligning this long-term view with more short term-oriented decisions and structures. This raises questions such as: How can long-term sustainability goals be aligned with short-term incentives? What are the appropriate governance structures to promote long-term thinking? What kind of reporting can align quarterly disclosures and financial markets with the necessary investments in sustainability transformation? How can path dependencies be broken (e.g., when retrofitting existing infrastructure, such as old factories) while ensuring profitability? What kind of change management is needed to align the transformation of business models, corporate processes, and individual competencies?

How Do You Align an Integrated Sustainability Strategy Across Fragmented Markets?

While many sustainability challenges are global in nature, market expectations and the regulatory requirements to address them differ from region to region. At the same time, multinational companies that operate in some or all of these regions face the challenge of formulating an integrated strategy that addresses this diversity while maintaining internal consistency. This raises questions such as: How will external sustainability requirements diverge or converge over time? How can companies align a global strategy with a fragmented regulatory and market landscape? How can the diversity of different strategy contexts be used as a source of experimentation, innovation, and scaling?

How Do You Align Your Sustainability Strategy with Your Value Chain and Other Business Actors?

Sustainability is a race that no company can win alone. For example, decarbonizing a product footprint requires collaboration across the entire value chain. Similarly, improving the working conditions of raw material suppliers, such as in the case of cobalt mines, is a challenge that transcends a single industry and benefits from the cooperation of different actors. In this context, integrating sustainability into strategy often requires working with other firms, including competitors, to engage in "co-opetition" (Brandenburger & Nalebuff, 2021). This raises questions such as: How can companies collaborate with competitors on pre-competitive sustainability issues? How can collaborative strategies be reconciled with the need to respect antitrust rules? How can novel forms of antitrust policies foster sustainability cooperation?

How can pre-competitive strategies be aligned with companies' search for individual competitive advantage? How can collaboration with non-industry partners foster competitive advantage? What are appropriate criteria for measuring and monitoring the success of sustainability partnerships?

How Do You Align Sustainability Strategy with Digital Transformation?
Digital transformation is a megatrend with enormous relevance for an integrated sustainability strategy. Sustainability requires the generation and analysis of new types of data, such as real-time carbon product footprints. In addition, to drive sustainability across the entire value chain, data must be shared across business partners. This raises questions such as: How can digitization increase the transparency and reliability of environmental and social performance data? What forms of data exchange are appropriate to make information accessible across the value chain? What are the incentives for data sharing while addressing security and privacy concerns? What role can digital industry data platforms play in reducing transaction costs and improving data quality? How can digitization engage previously silent stakeholders (e.g., by giving voice to workers or communities) in the supply chain? How can companies create a competitive advantage through digital platform solutions for sustainability?

3.5 Conclusion

Integrating sustainability into strategy creates significant opportunities to transform companies into change agents for a decarbonized, circular, resilient, and more socially just economy. This integration offers ample opportunities for businesses and their future market success. Realizing this potential requires a systemic integration of sustainability throughout the strategy process. In this endeavor, sustainability is a moving target. Consequently, integrating sustainability into strategy is not a one-time decision. It is the first step on a continuous journey.

How can sustainability be integrated into corporate strategy? We would like to highlight five takeaways from this chapter that invite further discussion:

1. To survive and thrive in the marketplace over the long term, companies need to move from stand-alone sustainability strategies to integrated sustainability strategies that redefine a company's corporate strategy (where to compete) and business strategy (how to achieve competitive advantage).

2. The development of an integrated sustainability strategy can follow a four-step process: (1) environmental scanning to analyze the external and internal environment; (2) strategy formulation as a multi-stage process with the formulation of strategic goals and strategic plans as core aspects; (3) implementation through cross-functional collaboration; and (4) evaluation with an integrated reporting approach.
3. The primary contribution of companies to society is to create value for their stakeholders. Here, customers are the lever that keeps a company profitable and serves as the foundation for everything else. Therefore, a company's strategy discussion must never lose sight of the customer.
4. Sustainability is a race that cannot be won alone. For this reason, becoming a truly sustainable business starts with proper internal and external collaboration, as The Road to Net Zero requires changes in a whole ecosystem.
5. Based on a proactive response to the VUCA world, an integrated sustainability strategy can deliver multiple business benefits and lead to technology and process optimizations that result in cost savings, improved performance, and increased resilience.

On the Road to Net Zero, however, a company's strategy journey matters not only to the firm and its investors, but also to other stakeholders, including nature and future generations. Therefore, creating transparency about a company's sustainability ambitions becomes increasingly important, as do the results achieved. For this reason, the next chapter, Chap. 4, focuses on *The Future of Corporate Disclosure*.

References

Andersen, I., Ishii, N., Brooks, T., Cummis, C., Fonseca, G., Hillers, A., Macfarlane, N., Nakicenovic, N., Moss, K., Rockström, J., Steer, A., Waughray, D., & Zimm, C. (2021). Defining 'science-based targets'. *National Science Review, 8*(7), nwaa186. https://doi.org/10.1093/nsr/nwaa186

Baier, C., Beckmann, M., & Heidingsfelder, J. (2020). Hidden allies for value chain responsibility? A system theory perspective on aligning sustainable supply chain management and trade compliance. *International Journal of Physical Distribution & Logistics Management, 50*(4), 439–456. https://doi.org/10.1108/IJPDLM-02-2019-0037

Bansal, P., & Roth, K. (2000). Why companies go green: A model of ecological responsiveness. *Academy of Management Journal, 43*(4), 717–736. https://doi.org/10.2307/1556363

Barney, J. B. (1997). *Gaining and sustaining competitive advantage*. Addison-Wesley.

Baumgartner, R. J., & Ebner, D. (2010). Corporate sustainability strategies: sustainability profiles and maturity levels. *Sustainable Development, 18*(2), 76–89. https://doi.org/10.1002/sd.447

Beckmann, M., Hielscher, S., & Pies, I. (2014). Commitment strategies for sustainability: How business firms can transform trade-offs into win–win outcomes. *Business Strategy and the Environment, 23*(1), 18–37.

Beckmann, M., & Schaltegger, S. (2021). Sustainability in business: Integrated management of value creation and disvalue mitigation. In R. J. Aldag (Ed.), *Oxford research encyclopedia of business and management*. Oxford University Press. https://doi.org/10.1093/acrefore/9780190224851.013.322

Bennett, N., & Lemoine, J. (2014). What VUCA really means for you. *Harvard Business Review, 92*(1/2) https://hbr.org/2014/01/what-vuca-really-means-for-you

Bjørn, A., Tilsted, J. P., Addas, A., & Lloyd, S. M. (2022). Can science-based targets make the private sector Paris-aligned? A review of the emerging evidence. *Current Climate Change Reports, 8*(2), 53–69. https://doi.org/10.1007/s40641-022-00182-w

BMW. (2021). How the BMW Group is increasing the pace to combat climate change. Retrieved April 20, 2023, from https://www.bmwgroup.com/en/news/general/2021/bmwgroup-sustainability.html

BMW. (2022). The BMW Group sustainability strategy rests on these six pillars. Retrieved April 20, 2023, from https://www.bmwgroup.com/en/news/general/2022/Sustainability360.html

Brandenburger, A., & Nalebuff, B. (2021). The rules of co-opetition. *Harvard Business Review, 99*(1), 48–57.

Burritt, R. L., Christ, K. L., Rammal, H. G., & Schaltegger, S. (2020). Multinational enterprise strategies for addressing sustainability: the need for consolidation. *Journal of Business Ethics, 164*(2), 389–410. https://doi.org/10.1007/s10551-018-4066-0

Bylund, P. L., & McCaffrey, M. (2017). A theory of entrepreneurship and institutional uncertainty. *Journal of Business Venturing, 32*(5), 461–475. https://doi.org/10.1016/j.jbusvent.2017.05.006

Chandler, A. D. (1962). *Strategy and structure: Chapters in the history of the industrial enterprise*. MIT Press.

Churet, C., & Eccles, R. G. (2014). Integrated reporting, quality of management, and financial performance. *Journal of Applied Corporate Finance, 26*(1), 56–64. https://doi.org/10.1111/jacf.12054

David, F. R. (2011). *Strategic management: Concepts and cases*. Pearson.

Dyllick, T., Belz, F., & Schneidewind, U. (1997). *Ökologie und Wettbewerbsfähigkeit*. Verl. Neue Zürcher Zeitung; Hanser.

Ellen MacArthur Foundation. (2013). Towards the circular economy: Economic and business rationale for an accelerated transition. Retrieved May 23, 2023, from https://ellenmacarthurfoundation.org/towards-the-circular-economy-vol-1-an-economic-and-business-rationale-for-an

Engert, S., Rauter, R., & Baumgartner, R. J. (2016). Exploring the integration of corporate sustainability into strategic management: a literature review. *Journal of Cleaner Production, 112*, 2833–2850. https://doi.org/10.1016/j.jclepro.2015.08.031

European Parliament. (2023). EU ban on sale of new petrol and diesel cars from 2035 explained. Retrieved April 21, 2023, from https://www.europarl.europa.eu/news/en/headlines/economy/20221019STO44572/eu-ban-on-sale-of-new-petrol-and-diesel-cars-from-2035-explained

Garst, J., Maas, K., & Suijs, J. (2022). Materiality assessment is an art, not a science: Selecting ESG topics for sustainability reports. *California Management Review, 65*(1), 64–90. https://doi.org/10.1177/00081256221120692

Hahn, R. (2013). ISO 26000 and the standardization of strategic management processes for sustainability and corporate social responsibility. *Business Strategy and the Environment, 22*(7), 442–455. https://doi.org/10.1002/bse.1751

Higgins, C., Stubbs, W., Tweedie, D., & McCallum, G. (2019). Journey or toolbox? Integrated reporting and processes of organisational change. *Accounting, Auditing & Accountability Journal, 32*(6), 1662–1689. https://doi.org/10.1108/AAAJ-10-2018-3696

Kaplan, R. S., & Ramanna, K. (2021). Accounting for climate change. *Harvard Business Review, 99*(6), 120–131.

Kuh, T., Shepley, A., Bala, G., & Flowers, M. (2020). Dynamic materiality: measuring what matters. Retrieved April 27, 2023, from https://doi.org/10.2139/ssrn.3521035

Mckeown, M. (2016). *The strategy book* (2nd ed.). Pearson.

Meffert, H., & Kirchgeorg, M. (1998). *Marktorientiertes Umweltmanagement: Grundlagen und fallstudien*. C. E. Poeschel.

Mintzberg, H., & Waters, J. A. (1985). Of strategies, deliberate and emergent. *Strategic Management Journal, 6*(3), 257–272. https://doi.org/10.1002/smj.4250060306

Oertwig, N., Galeitzke, M., Schmieg, H.-G., Kohl, H., Jochem, R., Orth, R., & Knothe, T. (2017). Integration of sustainability into the corporate strategy. In R. Stark, G. Seliger, & J. Bonvoisin (Eds.), *Sustainable manufacturing* (pp. 175–200). Springer. https://doi.org/10.1007/978-3-319-48514-0_12

Porter, M. E. (1979). How competitive forces shape strategy. *Harvard Business Review, 57*(2), 137–145.

Porter, M. E. (2008). The five competitive forces that shape strategy. *Harvard Business Review, 86*(1), 78–93.

Rajagopalan, N., Rasheed, A. M. A., & Datta, D. K. (1993). Strategic decision processes: critical review and future directions. *Journal of Management, 19*(2), 349–384. https://doi.org/10.1177/014920639301900207

Rockström, J., Steffen, W., Noone, K., Persson, A., Chapin, F. S., Lambin, E. F., Lenton, T. M., Scheffer, M., Folke, C., Schellnhuber, H. J., Nykvist, B., de Wit, C. A., Hughes, T., van der Leeuw, S., Rodhe, H., Sörlin, S., Snyder, P. K., Costanza, R., Svedin, U., … Foley, J. A. (2009). A safe operating space for humanity. *Nature, 461*(7263), 472–475. https://doi.org/10.1038/461472a

Salzmann, O., Ionescu-somers, A., & Steger, U. (2005). The business case for corporate sustainability. *European Management Journal, 23*(1), 27–36. https://doi.org/10.1016/j.emj.2004.12.007

Schmid, M. (2020). Challenges to the European automotive industry in securing critical raw materials for electric mobility: the case of rare earths. *Mineralogical Magazine, 84*(1), 5–17. https://doi.org/10.1180/mgm.2020.9

Steffen, W., Richardson, K., Rockström, J., Cornell, S. E., Fetzer, I., Bennett, E. M., Biggs, R., Carpenter, S. R., de Vries, W., de Wit, C. A., Folke, C., Gerten, D., Heinke, J., Mace, G. M., Persson, L. M., Ramanathan, V., Reyers, B., & Sörlin, S. (2015). Sustainability. Planetary boundaries: guiding human development on a changing planet. *Science, 347*(6223), 1259855. https://doi.org/10.1126/science.1259855

United Nations Statistics Division. (2022). SDG Indicators: Global indicator framework for the sustainable development goals and targets of the 2030 Agenda for Sustainable Development. Retrieved April 21, 2023, from https://unstats.un.org/sdgs/indicators/indicators-list/

van Marrewijk, M., & Werre, M. (2003). Multiple levels of corporate sustainability. *Journal of Business Ethics, 44*(2), 107–119. https://doi.org/10.1023/A:1023383229086

Van Zanten, J. A., & Van Tulder, R. (2021). Analyzing companies' interactions with the Sustainable Development Goals through network analysis: Four corporate sustainability imperatives. *Business Strategy and the Environment, 30*(5), 2396–2420. https://doi.org/10.1002/bse.2753

Verrier, B., Rose, B., & Caillaud, E. (2016). Lean and Green strategy: the Lean and Green House and maturity deployment model. *Journal of Cleaner Production, 116*, 150–156. https://doi.org/10.1016/j.jclepro.2015.12.022

Waldman, D. A., Ramirez, G. G., House, R. J., & Puranam, P. (2001). Does leadership matter? CEO leadership attributes and profitability under conditions of perceived environmental uncertainty. *Academy of Management Journal, 44*(1), 134–143. https://doi.org/10.2307/3069341

Wheelen, T. L., Hunger, J. D., Hoffman, A. N., & Bamford, C. E. (2017). *Strategic management and business policy: Globalization, innovation, and sustainability* (15th ed.). Pearson.

Whitehead, J. (2017). Prioritizing sustainability indicators: Using materiality analysis to guide sustainability assessment and strategy. *Business Strategy and the Environment, 26*(3), 399–412. https://doi.org/10.1002/bse.1928

Open Access This chapter is licensed under the terms of the Creative Commons Attribution 4.0 International License (http://creativecommons.org/licenses/by/4.0/), which permits use, sharing, adaptation, distribution and reproduction in any medium or format, as long as you give appropriate credit to the original author(s) and the source, provide a link to the Creative Commons license and indicate if changes were made.

The images or other third party material in this chapter are included in the chapter's Creative Commons license, unless indicated otherwise in a credit line to the material. If material is not included in the chapter's Creative Commons license and your intended use is not permitted by statutory regulation or exceeds the permitted use, you will need to obtain permission directly from the copyright holder.

4

The Future of Corporate Disclosure

Non-financial KPIs, Sustainability and Integrated Reporting

Thomas Fischer, Jennifer Adolph, Markus Schober, Jonathan Townend, and Oliver Zipse

4.1 Introduction

Crafting integrated strategies and pursuing integrated thinking (cf. Chap. 3) require new approaches to corporate reporting in order to provide stakeholders with relevant information about a company's business activities. In general, corporate disclosures should fulfil the information needs of stakeholders, minimise information asymmetries and enable better investment decisions by investors, thereby leading to a more sustainable allocation of capital (cf. Bonsón & Bednárová, 2015; Cormier et al., 2005; Deegan, 2002; Moratis, 2018). Corporate reports are also an important tool for communicating and internally steering the implementation of business strategies.

The development of new digital technologies and the rapidly changing business environment further influence corporate reporting. These trends may shift stakeholder expectations; therefore, companies will be challenged to meet the changing demands on corporate reporting (Barrantes et al., 2022). At the same time, the reporting and disclosure of intangible assets are

---

T. Fischer (✉) • J. Adolph
FAU Erlangen-Nürnberg, Nuremberg, Germany
e-mail: wiso-controlling@fau.de

M. Schober
FAU Erlangen-Nürnberg, Erlangen, Germany

J. Townend • O. Zipse
BMW AG, Munich, Germany

becoming increasingly important (Eierle & Kasischke, 2023; Haller & Fischer, 2023). In an environment characterised by high levels of volatility, uncertainty, complexity and ambiguity (VUCA), companies face various changes, such as the emergence of new business models. To adequately assess their situation, companies need to measure their ability to deal with these challenges (Duchek, 2020).

Another development relevant to the area of corporate reporting is the emergence of new regulatory requirements. To address the major challenges of climate change at a societal level, the European Union announced its EU Green Deal, a comprehensive set of plans that represent the need for the transformation of the economy and society (Council of the European Union, 2022). As part of the EU Green Deal, new sustainability reporting requirements have emerged for companies (e.g. EU Taxonomy, Corporate Sustainability Reporting Directive [CSRD]). To respond to these new requirements, companies can apply various strategies.

As described in Chap. 3, *Crafting Corporate Sustainability Strategy*, to stay on track on the Road to Net Zero, companies need to implement effectively integrated strategies that place sustainability at the heart of their corporate and business strategy. In addition to information on how environmental factors may affect a company's business activities, the need to provide reliable information on the environmental and social impacts and risks of business activities has become a major trigger in developing reporting standards in the EU. Thus, Chap. 4 is dedicated to new ways of reporting and leads into Chap. 5, *Creating Sustainable Products*, which looks in more detail at the transformation of the actual business operations.

The main objective of the remainder of this chapter is to provide an overview of how new forms of reporting have evolved over the last two decades, the reasons for this development and the implications for practice in implementing the new reporting standards. Section 4.2 shows the evolution from voluntary to mandatory sustainability reporting standards and from separate sustainability reporting frameworks to a combination of sustainability and financial reporting in an integrated report. Section 4.3 outlines the current legal regulations in the EU regarding non-financial reporting. The illustrative example of the BMW Group in Sect. 4.4 demonstrates the transition from the prior generation of separate reports to today's integrated sustainability reporting. The expert discussion between Prof. Oliver Zipse, Chairman of the Board of Management of BMW AG, Jonathan Townend, BMW Group's Head of Accounting and Prof. Dr Thomas M. Fischer, Chair of Accounting and Controlling at FAU Erlangen-Nürnberg, focuses on the practical challenges of integrated reporting (Sect. 4.5). Section 4.6 discusses current and future reporting challenges arising from the present and diversifying regulatory environment. Finally, the conclusion in Sect. 4.7 summarises the key takeaways from the chapter.

4.2 New Ways of Reporting

In the context of reporting, the definition of the scope of relevant information has changed significantly over the years. Before the 1970s, corporate reporting focused on a company's financial performance (Navarrete-Oyarce et al., 2021). As investors want to make their decisions based on reliable corporate information, financial reporting has been regulated early on by national governments to ensure this reliability and comparability. Following the United Nations (UN) Brundtland Report in 1987 (Brundtland, 1987) and the introduction of the 'Triple Bottom Line' (TBL) concept by John Elkington in 1994, the scope of corporate reporting began to broaden. In addition to financial information, environmental, social and governance (ESG) aspects of corporate activities became increasingly important as more investors considered these factors when evaluating a company (Alniacik et al., 2011; Böcking & Althoff, 2017). This has created a need for frameworks and standards that can be used to incorporate ESG issues into corporate disclosures in a concise and practical manner.

In 1997, the Global Reporting Initiative (GRI) was founded initiated by a multi-stakeholder initiative of companies, NGOs, audit firms, governments and others. Its aim was to develop an easy-to-use and standardised reporting framework that integrates economic, environmental and social aspects to enable informed decision-making by establishing specific metrics for sustainability issues (GRI, 2022). The GRI released its first global sustainability reporting guidelines in 2000 (GRI, 2022; Rowbottom & Locke, 2016). While only a few companies were listed in the GRI reporting database in the early years, companies increasingly adopted the framework with each revision of the guidelines in 2002 (G2), 2006 (G3), 2013 (G4) and 2015 (GRI Standards). Today, GRI is a globally disseminated framework and is recognised as a mature voluntary reporting standard (Chersan, 2016; GRI, 2022). Early on, GRI joined forces with international institutions such as the OECD, the UN Environment Programme and the UN Global Compact, a strategy that contributed to its success. For some time, however, GRI was considered difficult to compare with conventional financial reporting and therefore less investor-friendly, as the GRI framework was designed for a broader group of stakeholders than investors, such as society, employees, government or the media. In 2009, GRI announced that it would adjust its stakeholder focus to better meet the information needs of investors (Eccles et al., 2010; Rowbottom & Locke, 2016), creating a more integrated and comprehensive view of reporting.

The rationale for the GRI's revised audience was that other emerging and competing reporting guidelines focus on investors as a company's key stakeholder group. One example is the Carbon Disclosure Project (CDP), which was established in 2000 to transform environmental reporting by making climate-related environmental impacts measurable. By developing an independent scoring methodology, CDP measures the progress of companies or cities in their climate action and transparency. Each year, a ranking is produced and made available to the public. In 2022, CDP assessed 15,000 companies and claims to operate the world's largest environmental database (CDP, 2022).

Another reporting initiative was the British Accounting for Sustainability (A4S) Project, initiated by the former Prince of Wales in 2004. The so-called Connected Reporting Framework developed by the A4S differed significantly from the GRI in that it combined the financial indicators with considerations on sustainable corporate governance and linked these to a company's strategy and risk assessment (Druckmann & Freis, 2010; Rowbottom & Locke, 2016). South Africa has been equally important in the development of an integrated reporting format for financial and non-financial (i.e. sustainability-oriented) information. The first standard for an integrated report was introduced in 2002 with the King II governance code, which became mandatory for all companies listed on the Johannesburg Stock Exchange in 2010. This was the first time in the world that a regulatory body decided that integrated reporting should replace the previously often separate disclosure of financial and sustainability information (Brady & Baraka, 2013; Rowbottom & Locke, 2016).

There are many reasons for regulating sustainability-related disclosures. For example, mandatory non-financial disclosure can reduce information asymmetry if organisations present a holistic, and therefore more realistic and more complete picture of their different areas of organisational performance (Cormier et al., 2005; Du et al., 2010). In this way, non-financial reporting becomes an effective tool for legitimising an organisation's activities towards its stakeholders and society (Bonsón & Bednárová, 2015; Deegan, 2002; Lock & Seele, 2015). However, this information needs to be credible, as sustainability information is easily at risk of being discredited as 'greenwashing' or 'information overflow' (Marquis et al., 2016; Velte & Stawinoga, 2017). If done properly, sustainability reporting can be beneficial to the organisation itself, as it increases organisational transparency and contributes to organisational development (Diehl & Knauß, 2018).

Driven by the economic and financial crisis in 2008, various initiatives in the US and the EU started to consider new regulations to make reporting more comprehensive and integrated. However, the plethora of reporting frameworks available globally posed a challenge to the goal of improving regulatory requirements, as each framework had its own philosophy and focus. Despite the lack of international recognition of the Connected Reporting Framework, its representative, the former Prince of Wales, was able to launch a multi-stakeholder initiative at the annual A4S Forum, where companies, standard setters, UN representatives, investors and audit organisations jointly discussed a new internationally accepted reporting framework. As the parties involved agreed that sustainable management at the corporate level requires the combination of financial and non-financial reporting, a joint body—the International Integrated Reporting Council (IIRC)—was finally established in 2010 with the intention of developing the Integrated Reporting Framework (IRF) (Druckmann & Freis, 2010; Rowbottom & Locke, 2016). As of August 2022, the International Financial Reporting Standards (IFRS) Foundation assumed responsibility for the IRF (cf. IFRS Foundation, 2022b, p. 2).

The aim of the IRF is to guide organisations in the preparation of an integrated report (IR) as a new, comprehensive reporting format (cf. IFRS Foundation, 2022b, p. 2). The IRF focuses on a company's stakeholders, with particular attention given to investors and creditors to enable a more efficient and productive allocation of financial capital (IFRS Foundation, 2022b). Different types of capital are considered, such as financial, manufactured, intellectual, human, social, relationship and natural capital (IFRS Foundation, 2022b). This approach enables integrated thinking and business actions focused on long-term value-creation interdependencies (IFRS Foundation, 2022b). Companies can use the IRF to clearly communicate how their business activities lead to the creation, preservation or erosion of value over time, taking either a short-, medium- or long-term perspective. Figure 4.1 illustrates how the BMW Group applies this (reporting) process of transforming value from input capitals into output capitals.

The IRF proposes seven guiding principles that form the basis for the preparation and presentation of integrated reports (IFRS Foundation, 2022b). These relate to the strategic focus and future orientation of the report, the connectivity of information, the management of stakeholder relationships, the principle of materiality, the reliability and completeness of the report and its consistency and comparability (IFRS Foundation, 2022b). With regard to the content of the reporting, the IRF defines eight interrelated elements: an

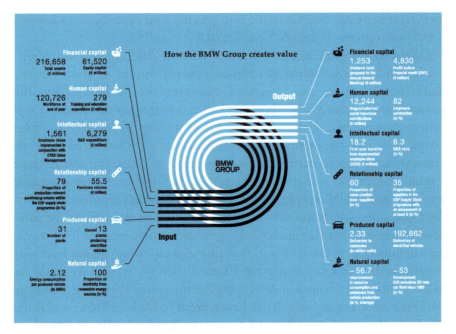

Fig. 4.1 Elements in an integrated report to explain a firm's value creation, preservation or erosion as applied by the BMW Group (BMW AG) (2021, p. 57)

overview of the organisation and its embeddedness in the external environment, governance structures, a description of the business model, the risks and opportunities of the business operations, the strategy and resource allocation to achieve the organisation's objectives, the organisation's performance, an outlook on future challenges and the basis for preparation and presentation (IFRS Foundation, 2022b).

To date, mandatory use of the IRF is required in South Africa and Japan, but it has not become the global industry standard (Threlfall et al., 2020, p. 21). In the European Union in particular, the legislative landscape imposes different legal requirements on its member states, as the next section illustrates.

4.3 The Current Legislative Landscape in the EU

In the EU, some countries had implemented mandatory reporting standards for social and environmental aspects early on, such as France in 2001 or the Scandinavian countries in the 1990s (Hess, 2007; Hoffmann et al., 2018). In 2014, the EU Non-Financial Reporting Directive (NFRD) introduced mandatory reporting requirements with the aim of promoting harmonisation, transparency and comparability (Directive 2014/95/EU, see European

Parliament and the Council of the European Union, 2014). Since fiscal year 2017, certain large companies in EU member states have been required to provide additional 'non-financial information' in their annual disclosures. In its most basic sense, non-financial reporting focuses on data other than financial data (Baumüller & Schaffhauser-Linzatti, 2018; Loew & Braun, 2018). In detail, however, the relevant content relates to environmental, social and governance aspects, often referred to as 'ESG factors' (cf. Baumüller & Schaffhauser-Linzatti, 2018; Loew & Braun, 2018). These ESG factors have gained particular importance since the financial crisis in 2007/08, especially in the financial industry. Indeed, there are even official recommendations on ESG criteria for financial products, such as the Statement on Disclosure of ESG Matters by Issuers of the International Organization of Securities Commissions (IOSCO, 2019) or rankings based on ESG criteria, such as the Fitch Ratings ESG Relevance Score (Fitch Ratings, 2019). According to a study by Union Investment (2021), 78% of all large-scale investors in Germany consider sustainability issues in their investment decisions.

To steer investment towards sustainable business models at the regulatory level, the EU taxonomy for sustainable activities initially set requirements for large listed companies with more than 500 employees, which are obliged to disclose non-financial information under Article 19a or 29a of the Directive 2013/34/EU (Böcking & Althoff, 2017). As the European Commission has pointed out, the 'disclosure of non-financial information is vital for managing change towards a sustainable global economy by combining long-term profitability with social justice and environmental protection. In this context, disclosure of non-financial information helps the measuring, monitoring and managing of undertakings' performance and their impact on society' (Directive 2014/95/EU, L 330/1). However, the legislation itself was criticised by public and private actors for still leaving much room to manoeuvre and interpretation, thus weakening the intended comparability and credibility (cf. Baumüller & Schaffhauser-Linzatti, 2018; Global Compact Network Deutschland e.V. & econsense, 2018; Loew & Braun, 2018; Velte, 2017). In parallel, another point from the EU Action Plan for financing sustainable growth, 'Strengthening sustainability disclosure and accounting rule-making' (European Commission, 2023), was addressed by a revision of the mandatory reporting approach by the European Commission and resulted in the EU CSRD (Directive 2022/2464/EU) in 2022 (European Parliament and the Council of the European Union, 2022).

The revised mandatory reporting legislation will change the future scope of the required disclosures for preparers and users. The CSRD regulations will be applied in four stages (Directive 2022/2464/EU, see European Parliament and the Council of the European Union, 2022) and the CSRD is expected to apply to 50,000 companies (European Commission, 2022).

In terms of the content of CSRD disclosures, companies will be required to report information on environmental (E), social (S) and governance (G) issues regarding several pre-defined subtopics, such as climate change mitigation, workforce or business ethics and corporate culture (Directive 2022/2464/EU, see European Parliament and the Council of the European Union, 2022). Information on these aspects must be disclosed if it meets the principle of 'double materiality'. This requires companies to report information necessary to understand (1) the company's impact on sustainability matters ('inside-out' perspective) *and/or* (2) how sustainability matters affect the company's business development, performance and position ('outside-in' perspective). More precisely, either one or both of these conditions must be met for sustainability aspects to be reported under the CSRD, thereby broadening the scope of reporting content.

With respect to those ESG issues deemed material, companies are required to disclose information on (1) the business model and strategy; (2) time-bound sustainability targets and GHG reduction plans, including progress in each reporting year; (3) the role of the supervisory and management bodies with respect to sustainability aspects, including incentive schemes; (4) policies and due diligence processes, including the results of these policies; (5) principal risks and how they are managed; and (6) performance indicators relevant for disclosure. Furthermore, Article 19a (2) of Directive 2013/34/EU is expanded, for example, by introducing the new term 'resilience' and requiring companies to disclose related information (Directive 2022/2464/EU, L322/24, see European Parliament and the Council of the European Union, 2022). Another change in reporting requirements concerns information on intangible assets, which have become an important driver of company value (Fischer & Baumgartner, 2021).

Independent of the new EU regulations, other standards and frameworks for sustainability disclosure continue to exist. Some frameworks, such as the GRI, IRF or CDP, are explicitly mentioned in the CSRD to 'minimise disruption' to companies (Directive 2022/2464/EU, L 322/29, see European Parliament and the Council of the European Union, 2022). In 2021, the European Commission appointed the European Financial Reporting Advisory Group (EFRAG) as a technical adviser for the development of the European Sustainability Reporting Standards (ESRS). The ESRS will follow a modular structure and will be divided into cross-cutting standards and topical standards (representing the ESG topics) (EFRAG, 2022). The first set of the ESRS draft contains a total of 12 modules with 82 disclosure requirements and specified data points (equivalent to Key Performance Indicator [KPIs]). Two modules are available for cross-cutting standards (ESRS 1 + ESRS 2), five modules for environmental topics (ESRS E1–ESRS E5), four modules for social topics (ESRS S1–ESRS S4) and one module for governance topics (ESRS G1) (EFRAG, 2022). The ESRS are intended to become the primary and partially binding framework for reporting under the CSRD.[1]

Another change introduced by the CSRD compared to the NFRD is the requirement to publish sustainability information electronically in accordance with the European Single Electronic Format, which has been applicable for financial information since 2020 (Directive 2022/2464/EU, see European Parliament and the Council of the European Union, 2022). Furthermore, sustainability information must be part of the management report and be subject to limited assurance by external and independent auditors. Having sustainability-related information prominently displayed in the management report as one of the first chapters of each financial report takes us a step closer to marrying non-financial (sustainability) information and financial information. As highlighted by the EU Parliament in a press release, the CSRD is a milestone as '[f]inancial and sustainability reporting will be on an equal footing […] [to enable better] comparable and reliable data' (European Parliament, 2022).

The complexity of reporting in the EU increased further in 2020 with the implementation of the EU taxonomy, which is applied by companies in their reporting starting for the fiscal year 2021 (Regulation (EU) 2020/852, see European Parliament and the Council of the European Union, 2020). Taxonomy is a classification system for sustainable economic activities as one tool of the EU Action Plan on Financing Sustainable Growth (2018) and the EU Green Deal (2019). It is intended to provide a frame of reference for investors and companies that recognises corporate activities as environmentally sustainable if they make a substantial contribution to at least one of six environmental objectives of the EU taxonomy: (1) climate change mitigation, (2) climate change adaptation; (3) sustainable use and protection of water/marine resources; (4) transition to a circular economy; (5) pollution prevention and control; or (6) protection and restoration of biodiversity and ecosystems. The EU taxonomy is intended to be a 'transparency tool' (European Commission, 2021, p. 1), as it aligns the financial value of a firm's corporate activities with reporting on specific environmental criteria. However, the EU

---

[1] The Commission has adopted the Delegated Act on the first set of European Sustainability Reporting Standards (ESRS) for use by all companies subject to the Corporate Sustainability Reporting Directive on July 31st 2023 (EFRAG, 2023; EU Commission, 2023).

taxonomy is expected to be revised over time to include other economic sectors that are currently outside its scope (European Commission, 2021). As a result, the regulatory requirements for corporate sustainability reporting will continue to change and expand.

In practice, the merging of financial and non-financial reporting has not taken place from 1 year to the next, but is the result of a longer period of transformation, as the following case of BMW illustrates.

4.4 Integrated Reporting in Practice

Companies that started to voluntarily apply sustainability-related frameworks early to extend their mandatorily disclosed financial information have achieved a good starting position to launch integrated reporting. This becomes obvious in the case of BMW, which has continuously developed its sustainability reporting since the 1970s and made it part of the strategy process.

In 1973, BMW became the first company in the automotive industry worldwide to appoint an environmental protection officer. After the turn of the millennium, the first sustainability report, the 'Sustainable Value Report', was introduced for the fiscal years 2001 and 2002. Even before the introduction of that report, BMW had already published reports on the environmental impacts of its operations and the measures taken to counteract them. When BMW began publishing its voluntary Sustainable Value Report, it initially did so on a bi-annual basis and, subsequently, beginning in 2012, on an annual basis (Value Reporting Foundation, 2022a). While the first reports did not follow specific reporting standards, the company has, since 2005, adopted the GRI standard for sustainability reporting and has voluntarily committed to the highest GRI application level ('comprehensive option') since 2008 (BMW AG, 2021). Considering that, according to a CSR-reporting ranking in Germany, only three major German corporations (Daimler, BASF and BMW) have committed to the highest GRI application level in their sustainability reporting as of fiscal year 2020 (Institut für ökologische Wirtschaftsforschung and future e.V.—verantwortung unternehmen, 2022), this further supports the company's pioneering role in German industry and led to BMW being recognised by the Carbon Disclosure Project (CDP) in 2014 (BMW Group, 2014). With regard to auditing, BMW has strengthened its credibility by committing to a voluntary limited assurance audit of its sustainability report since 2013 (BMW Group, 2021). BMW claims to be the first premium car manufacturer in the world to finally complete the transition from separate sustainability reporting to a fully integrated report (BMW AG,

2021). BMW's integrated reports for the fiscal years 2020 and 2021 follow the voluntary framework of the International Integrated Reporting Council, the Integrated Reporting Framework (BMW AG, 2021, 2022b).

According to BMW's own statements in the context of a case study published by the Value Reporting Foundation (2022a), integrated reporting appears to be just another logical step in a continuous process of transformation across the entire company. The organisational perspective changes when sustainability becomes the core of corporate strategy. Then there no longer seems to be a need for a separate sustainability strategy (cf. Chap. 3), but sustainability becomes a central factor in corporate decision-making as part of an integrated corporate strategy. This gives rise to a new perspective on value creation, the so-called Integrated thinking, which views 'sustainability, social impact and economic and business success as mutually dependent' (Value Reporting Foundation, 2022a, p. 14).

In the following expert dialogue between Prof. Oliver Zipse, Chairman of the Board of Management of BMW AG, Jon Townend, BMW Group's Head of Accounting and Prof. Dr Thomas M. Fischer, Chair of Accounting and Controlling at FAU Erlangen-Nuremberg in Germany, the implications of implementing integrated reporting at BMW Group are discussed and reflected upon in more detail.

4.5 Expert Conversation on the Implementation of Integrated Reporting at BMW Group

Fischer: 'Integrated reporting' is a combination of mandatory financial and non-financial information as well as voluntarily selected non-financial factors to communicate a company's value creation potential in a concise manner. BMW has prepared an integrated report for the first time for fiscal year 2020. What prompted you to do this?

Zipse: If you're an entrepreneur, you have to build up trust and you have to make sure that the business works. There are two key ingredients. The first is: What gets measured, gets done—and that builds up trust. If you walk your talk, you set specific targets and measure them. Trust is even higher when it builds upon full transparency. The second ingredient is to make sure you set the right goals. An organisation needs direction and trust in its leadership. So we thought it was a good time to bring together financial targets and reporting together with our non-financial targets and reporting, which we have been doing for many years. It's not a one-off for us. We started more than 20 years ago to set a greater number of concrete goals, make them transparent and report on them to the outside world. We have had a sustainability report since 2001 reporting on a wide range of measures.

Fischer: I agree about the internal impact of reporting, but what about the external drivers? What drove your decision?

Zipse: Society and the political environment are changing in such a way that the credibility they expect from you is based on proven facts. We are a pioneer in merging our two reporting formats—non-financial and financial reports—and making integrated strategy and reporting a part of our internal policy. We want to bring up external transparency in line with our corporate strategy. To date, this has been a real success story. But bringing them together was a bold step. It sounds easy, but it's quite difficult. You have to be very precise about the quality of each measure. Since we have auditors, it is much stricter that what you report has to be correct. However, part of our philosophy has always been to take the next step, to act. Sustainability is becoming a cornerstone of our corporate strategy. It is no longer something you report on to look good. If you don't act sustainably in what you do, you will quickly disappear from the market.

Fischer: So instead of 'sheer driving pleasure', it's now 'sheer reporting pleasure', so to say?

Townend: [laughter] Not so long ago, we still had a very heterogeneous standard-setting world. Different standard setters, different focuses. Now, the world has changed enormously. We have seen, for the first time, the major standard setters outside the financial sector working together intensively. The EU has also taken up the issue of sustainability reporting. A major development at the end of 2021 was that the IFRS Foundation announced the formation of the International Sustainability Standards Board (ISSB), following strong market demand for its establishment. I think this development will give corporates a much clearer framework of what is expected of them and that it will ultimately lead them to what investors are looking for. Investors want to see companies set targets and report transparently on these targets because this makes their actions comparable with what other companies and competitors are doing.

Fischer: You mentioned that you want to improve trust in communication. This is not an easy step to take: Creating these new reporting processes is a complex task. What are the main challenges compared to the purely financial reporting of previous decades?

Townend: When it comes to non-financial reporting, you're dealing with a large number of players within the reporting process. So it's not just the accountants or the controllers and the euros. You're dealing with a wide range of numbers that are more technical. There are also interdependencies between these figures. If you look at a CO_2 figure in our non-financial reporting, it's not just one person sitting at a desk calculating the CO_2 figure. You've got to look at the cars we sell, the type of cars we sell. You've got links to the engineering department. A lot more players are involved, and it's important that every involved colleague knows what the other one is doing. You have to work as a team. Last year, we had a long discussion about the responsibility for non-financial reporting. You have to make sure that the dependencies and the responsibilities are 100% clear. It's about making sure that every single player on the team is running in the right direction and fully understands the implications of his/her role.

Fischer: And in this team play, who is driving the process? Is it still the CFO?

Townend: The CFO is driving the process, because what you realise is that we're dealing with figures. They may not be euros, but they're still figures. And if you look at which department within a company is really best placed to understand how numbers are consolidated, how an internal control system ensures the quality of those figures, it is the accounting department.

So, we have a very important role to play. But as I mentioned, the technical side of the figures—kilowatt-hours, CO_2 or other aspects of the whole process chain—really requires experts. Among others, we work closely with Thomas Becker (BMW VP Sustainability, Mobility) and his team on the environmental figures. And, of course, we work closely with our colleagues in HR on the social metrics, the diversity metrics and the training and other metrics, as well as with the legal department on the governance issues.

Fischer: You mentioned controlling issues and addressed them to a professor of controlling, so I always like it to get some references to my home turf [laughs]. However, at the end of the day, you have to come up with a profit figure—a return on investment. You have to pay dividends in the end. How difficult is it to select the right performance measures, the non-financial or the quantitative indicators, to explain the resulting financial performance?

Zipse: Our transition to integrated reporting was made in anticipation of something that we believe is going to happen anyway. Look at the supply chain legislation in Germany: the German Act on Corporate Due Diligence Obligations in Supply Chains (Lieferkettensorgfaltspflichtengesetz). Look at the [EU] Taxonomy and CSRD. They all have to do with transparency—transparency for society and investors. It becomes mandatory to demonstrate that you are on a continuous improvement path in whatever you do. It's not just about setting long-term targets. It's about getting better every year in everything you do: CO_2 emissions, water consumption, energy consumption, energy sources. What was the performance of our cars? How well trained are our employees? We want to see progress year on year. And this whole framework of integrated reporting is a good indicator that we are a good investment.

Fischer: You are right. That is what reporting is all about: informing about the development of the business. Do you think that investors appreciate the integrated non-financial information?

Zipse: Sustainability reporting is also an investment instrument. With integrated reporting, we can prove that we are not only a highly profitable but also a sustainable investment for the future. With our transparency, we have been able to demonstrate for many years that we are improving year on year on the most important environmental factors, such as energy consumption and all forms of resource use. For example, in 2021 and 2022, we significantly overachieved our CO_2 emission targets set by the European Union for our new car fleet.

That's just one factor that, if you don't make it transparent, the outside world, the investors, may not even know that we are better than we are actually required to be. You have to be an attractive investment, an attractive employer and an attractive carmaker for customers—and that is the whole framework in which we operate.

Fischer: So is integrated reporting then the end of accounting?

Townend: [laughs] No, integrated reporting is the future of the accountants. Maybe it's the end of controlling [laughing]. No, seriously, it's definitely not the end of accounting. I think that integrated reporting is an integration of non-financial and financial reporting. Financial reporting must and will remain important in the years to come because financial reporting is about the reliability and quality of the company's management. Management sets out to do something at the beginning of the year and reports on what it has done at the end of the year. How close it is to what was expected is an indication of how well the company has been steered and managed internally. The accounting policies can also tell you something about the management. Is it an aggressive management? Or one with more conservative accounting policies and more prudent management? And without cash flow, we can't invest in the future anyway. So, I don't think you're going to move away from financial reporting.

Fischer: Does the capital market perceive non-financial reporting in the same way as financial reporting?

Townend: Non-financial reporting is now evolving to be on equal footing with financial reporting from an investor perspective. And the expectation around non-financial figures is that they will be derived and prepared with the same due process, care and attention to quality that we know from financial reporting.

Fischer: That's a crucial aspect: I am providing additional information to the capital market or to other stakeholders in order to build or restore trust. But we also know from empirical accounting research that additional information can be an additional risk. A stakeholder might say, 'OK, now you are providing me with facts that I didn't even know about, and now I see it in combination with, for example, cash flow, and it doesn't always look like it's going to have a positive impact'. What do you do as a company in this situation to achieve the results you originally intended? In a volatile environment, it is then a challenge for management to act reliably in the long term and to avoid myopic behaviour.

Townend: You raise an interesting point about the number of KPIs to be reported. I strongly believe that it is better to report less than more, in line with the current legislation. They need to be clearly derived from our strategy. An integrated report shows what the company is doing and the 'why' behind it, and this needs to be derived from the integrated strategy. A link is also needed between strategy and the remuneration of the Board of Management. The risk of reporting too many figures is that you will end up with conflicts and figures that are difficult to interpret. It's no coincidence that when rating agencies look at the same set of non-financial figures, they come to completely different conclusions about whether a company is a good or a bad investment.

Fischer: So less is more?

Townend: It's very risky to start reporting too many figures because you might lose focus on what the company is aiming for. I'm very much in favour of principle-related guidelines because they give companies the flexibility that they need to differentiate themselves in their reporting. And I think that's what investors need to know. They need to get a feeling what's behind the figures. It's not just a box-ticking exercise. It's really something that is selling or reflecting the company.

Fischer: A division into business segments would probably make reporting even more complex. But you mentioned the automotive segment as a whole, which brings me to the next topic of discussion: the EU taxonomy.

Zipse: Good point! Let's move on. The taxonomy is, of course, an important piece of regulation coming from Brussels. What is your assessment? Is it helpful for the development of the industry as a whole?

Fischer: Well, whether it is helpful or not, I think that is still an open question and a debate that we cannot conclude here right at the moment. But it's a very important topic that you're raising, and one that has gained momentum over the last month or year. We now have different levers for reporting and also for disclosure when it comes to discussing the impact that a company has. We have a strictly microeconomic or even a segmental perspective in financial reporting and in the integrated report. And the issue that you have raised is more on a higher aggregated basis; so, for example, the discussion about the environmental or social footprint is not only done at the corporate level, but also at the sector level. And the interesting thing is that I also see emerging discussions—for example, among macroeconomic experts—that we need new KPIs, new metrics, to determine whether a business period was successful.

The acronym KPI, for example, is then translated as a 'Key Purpose Indicator'. There is an initiative, the Value Balancing Alliance, which says, 'OK, at the end of the day, you have the environmental footprint or the social footprint, perhaps in combination with cash flow or dividend payments', but then it is all about the transformation of resources and the process of creating value for the stakeholders. That is coming more and more into focus.

Zipse: Do you expect integrated reporting to become the standard for all industry players in the near future?

Fischer: I'm not in the political arena, but the ISSB's exposure draft of the practice statement for the management commentary, which was published in May 2021, is more or less written under the guideline of integrated reporting, even if I think they don't use the term. But it's implicit, so let's say that integrated reporting is the guideline for the future. In addition, most of the financial statements and therefore the management commentaries, especially in Europe, are already prepared according to the IFRS. So, at the end of the day, I would say that, not too long in the future, integrated reporting will become a very important format for corporate disclosures.

Zipse: We think so, too. In today's world, you can only be an entrepreneur if you take into account all the resources that you use—social resources, environmental resources, financial resources and natural resources—because they are scarce and limited. I think the whole world is coming to the conclusion that you cannot have a market economy if you do not pay for resources or at least if you are not transparent about the use of resources. We feel that this is going to happen very quickly.

Fischer: So, as you mentioned, it will be about using scarce resources as efficiently as possible, and then explaining the value creation process in the company more comprehensively than just going over the financial report. I am confident that perhaps, in the near future, we will be able to discuss the progress and the next integrated report that BMW will prepare.

4.6 The Future of Reporting: Opportunities, Challenges and the Role of Integrated Reporting

Together with an integrated strategy, the Integrated Reporting Framework is leading the way in the sustainability-driven business transformation of companies. This is illustrated by the BMW Group case presented in the previous sections.

As described in Sects. 4.2 and 4.3, the regulatory framework for reporting has evolved from voluntary sustainability reporting (GRI) and climate reporting (CDP) to integrated reporting (IR) and the new mandatory EU reporting framework (CSRD) and standards (ESRS), as well as the EU's classification system (EU taxonomy).

Navigating this dynamic regulatory landscape presents a number of opportunities and challenges, which are explored in the following section. Further, the future role of integrated reporting is discussed.

4.6.1 Opportunities of New Ways of Reporting

A major opportunity offered by the new ways of reporting is that they affect the process of developing an integrated strategy, integrated decision-making and management (cf. Chap. 3). This can be observed in both the Integrated Reporting Framework and the new EU reporting regulations.

The Integrated Reporting Framework has emerged as part of a management philosophy called Integrated Thinking (cf. IIRC, 2019; Value Reporting Foundation, 2022b).As a 'multi-capital management approach' (IIRC, 2019, p. 5), it thus pursues '[l]inking purpose and values to strategy, risks, opportunities, objectives, plans, metrics and incentives throughout the organization […] [to enable] better decision-making' (IIRC, 2019, p. 5). Therefore, integrated reporting, as one of the principles of Integrated Thinking, provides the opportunity to build a bridge between strategy and the assessment of sustainability performance that spans all areas of an organisation and consequently leads to integrated decision-making.

The impact of applying integrated thinking principles to strategy and reporting can be seen in the case of BMW Group. BMW Group's objective in adopting the integrated reporting format for its annual report was to provide a clear and comprehensive insight into the BMW Group and to explain the organisation's activities in a transparent, comprehensible and measurable way

(BMW AG, 2022a). With its integrated report, BMW Group explains its corporate strategy aimed at achieving both financial and non-financial targets (e.g. earnings before tax (EBT) margin, share of electrified cars in total deliveries and reduction of CO_2 emissions per vehicle produced) (BMW AG, 2022a). Thus, in the described case, the integrated report, on the one hand, serves to communicate the strategy internally and externally to diverse stakeholder groups. On the other hand, it serves as an internal management tool to monitor and control the achievement of objectives, and it can be used as a basis for informed decision-making in the company's strategy process.

Although the EU's new mandatory reporting framework (CSRD) and the development of European sustainability reporting standards (ESRS) are not based on any specific management philosophy, they have the potential to impact how companies communicate their strategies to stakeholders. The recent changes will also affect the company's decision-making processes and business activities. This can be exemplified by the following two aspects:

First, a change in responsibility and in the attention paid to sustainability matters is to be expected at the individual level among management executives and in bodies such as management or supervisory boards. In the past, sustainability reporting did not always receive the same level of attention from a company's management and board level as financial reporting attracted. This is expected to change with the introduction of the ESRS (cf. EFRAG, 2022), and will be in line with the basic idea of the CSRD in terms of aligning the relevance of financial reporting and sustainability reporting (Zülch et al., 2023).

The revised draft of the ESRS 2—General Disclosures, published in November 2022, not only covers firm disclosure on 'the elements of its strategy that relate to or affect sustainability matters, its business model(s) and its value chain' (EFRAG PTF-ESRS, 2022, p. 10), as well as reporting standards on the assessment of sustainability matters. It also names an obligation to provide information on the governance of this disclosure.

Regarding governance, it demands disclosure on 'whether, by whom and how frequently the administrative, management and supervisory bodies, including their relevant committees, are informed about material impacts, risks and opportunities [...], the implementation of sustainability due diligence and the results and effectiveness of policies, actions, metrics and targets adopted to address them', as well as how they 'consider impacts, risks and opportunities when overseeing the undertaking's strategy, its decisions on major transactions and its risk management policies' (EFRAG PTF-ESRS, 2022, p. 9). Further disclosure on whether 'incentive schemes are offered to members of the administrative, management and supervisory bodies that are linked to sustainability matters' (EFRAG PTF-ESRS, 2022, p. 9) also seems to have become part of the reporting standards. In addition, the ESRS draft states disclosure requirements on 'how the interests and views of its stakeholders are taken into account by the undertaking's strategy and business model(s)' (EFRAG PTF-ESRS, 2022, p. 12) to account for the aspect of impact.

On the one hand, these new disclosure standards can certainly be seen as a challenge with respect to their implementation. On the other hand, the increased responsibility of management and supervisory individuals and bodies can be regarded as an opportunity to raise awareness among the management about sustainability matters and the associated opportunities and risks.

Second, the new reporting regulations will further accelerate the transformation and governance of sustainability-driven business models. Although non-financial information cannot be directly expressed as a monetary value, it could affect how stakeholders perceive a company's financial performance over time (cf. Böcking & Althoff, 2017, p. 246). Sustainability aspects have an impact on an organisation's opportunities, risks and the future going concerns of its business model. Moreover, sustainable business development can be beneficial for organisational resilience (cf. Schmidt & Strenger, 2019, p. 483). Sustainability risks can increase reputational risks, as they are highly relevant to society and subject to various regulatory developments. Consequently, sustainability reporting on ESG issues can contribute to reputation risk management (Bebbington et al., 2008, p. 337ff.). Non-financial KPIs are therefore early risk indicators and should be considered in an organisation's strategy (cf. Böcking & Althoff, 2017, p. 249). Integrating sustainability considerations can ensure long-term profitability, thereby enhancing a company's shareholder value (cf. Schmidt & Strenger, 2019, p. 483). As a further implication of regulatory reporting requirements, mandatory sustainability disclosures increase compliance sensitivity (cf. Bachmann, 2018, p. 233).

In addition to the opportunities and potential for sustainability-oriented corporate development through new forms of reporting, operational and regulatory challenges remain that need to be addressed.

4.6.2 Challenges of New Ways of Reporting

One of these remaining challenges is the operational implementation of the new reporting framework and standards. A 'CSRD readiness' ranking analysed by Zülch et al. (2023), which takes into account 160 management and sustainability reports of companies listed in the DAX, MDAX and SDAX, supports the assumption that companies that have previously engaged in sustainability reporting on a voluntary basis are better prepared for the implementation of the new reporting requirements and standards in the EU. Most of the ten top-ranking companies apply several recognised international standards for sustainability reporting and have a sustainability report integrated into their management reports.

With regard to 'CSRD readiness', two groups of companies emerge. Those that have been less advanced in sustainability reporting will now be challenged to define and establish responsibilities, strategies and processes to implement the regulatory requirements and increase personnel capacity to do so. The other group of companies has already voluntarily implemented standards, perhaps even including an integrated reporting framework and audits. This second group of companies will have to consider how to deal with their advanced reporting formats in light of the new regulations, as the integration of the sustainability report into the management report under the CSRD seems to be of limited scope compared to the Integrated Reporting Framework (cf. Zülch et al., 2023). However, if organisations exclude specific, detailed, stakeholder-oriented sustainability information from their integrated report, they will face the question of where to publish this information. Barrantes et al. (2022) therefore expect that organisations will continue to use separate sustainability reports in the medium term, but will eventually find ways to restructure them and to increasingly link them to corporate reporting content (cf. Barrantes et al., 2022, p. 90).

Another important challenge is the reporting of 'key intangible resources' in the context of sustainability reporting, which is reflected in a recent publication by Haller and Fischer (2023). In conventional financial reporting, the discussion about reporting of intangible resources, such as data, reputation, brand names or relationships, has become increasingly important because intangible resources can have a direct monetary impact on a company's net worth, as well as a strategic, indirect impact on a company's future opportunities, risks and competitive advantages. Consequently, inadequate representation of intangible assets can lead to an information gap in the management report, leaving room for interpretation that could potentially create a gap between book value and market value.

This issue becomes even more relevant in the context of sustainability reporting, as sustainability issues are predominantly intangible in nature and can directly and indirectly affect a company's opportunities and risks to create, preserve or erode value. The CSRD therefore contains, for the first time in reporting history, a regulatory impulse to report on 'key intangible resources'. However, as Haller and Fischer (2023, p. 82) point out, in the

CSRD, the EU considers and regulates under the term 'key intangible resources' only those intangible resources on which a company is materially dependent as part of its value creation activities (outside-in perspective). According to Haller and Fischer (2023, p. 83), this understanding of 'key intangible resources' does not seem to be in line with the fundamental principle of double materiality on which the CSRD is based. Considering only reporting regulations on 'key intangible resources' that might affect the company's business development, performance and position (outside-in perspective) leaves in question how to deal with information on 'key intangible resources' that would impact the company's activities on sustainability issues (inside-out perspective). In addition, the CSRD does not provide a categorisation of 'key intangible resources'. Both aspects—the lack of a definition and a categorisation of intangibles for external reporting—will decrease the comparability of related corporate disclosures (Haller & Fischer, 2023).

4.6.3 The Future Potential of Integrated Reporting

In terms of reporting format, the introduction of the CSRD changes corporate reporting in the EU insofar as the CSRD intends that companies include sustainability information in the management report of the annual report (cf. Baumüller et al., 2021).

Although this first step towards integrated reporting does not seem to be comparable with an Integrated Reporting Standard under the IR Framework, the legislative development in the EU can be credited with a certain push towards integrated reporting (see Barrantes et al., 2022, p. 90). Furthermore, the International Sustainability Standards Board (ISSB) has committed to additional development of the IR framework towards an international corporate reporting framework (cf. IFRS Foundation, 2022a), which speaks for the future relevance of the IR framework. In order to ensure the future global recognition of different reporting frameworks, the CSRD already states that the ESRS to be developed shall be consistent with the future basic reporting standards of the ISSB (Directive (EU) 2022/2464, L 322/29, see European Parliament and the Council of the European Union, 2022). Whether the CSRD will be able to achieve its objectives remains to be seen, as does the role the GRI Standards, the Carbon Disclosure Project or the Integrated Reporting Framework will play alongside the ESRS in the future.

On the one hand, an international trend is evident towards greater harmonisation of reporting frameworks and standards, which could lead to more homogeneous reports. On the other hand, the possibilities offered by digitalisation are encouraging a trend towards customised reporting formats. As the main target group for reporting expands from shareholders to various other stakeholders, such as employees, NGOs or sustainability experts, different information needs are growing (Barrantes et al., 2022). Whether this will be met in the future by adapting the communication format of reporting, such as an online platform with a search function, or by maintaining the diversity of different reporting frameworks also remains an open question for the future.

4.7 Conclusion

Corporate reporting is currently evolving faster than ever before. While companies must satisfy the information needs of diverse stakeholders, including employees, customers, media or experts, they are required to adapt their reporting processes to meet new legal requirements, such as the CSRD or the EU taxonomy. In addition, the landscape of voluntary frameworks intended to strengthen integrated thinking is currently undergoing adjustments to enhance the comparability of disclosure globally.

On the regulatory side, the most fundamental change in the EU is the introduction of the CSRD, which, in contrast to the previous NFRD, integrates sustainability reporting as a mandatory part of the management report and imposes an audit requirement with limited assurance. In addition, the information to be reported on environmental (E), social (S) and governance (G) aspects must comply with the principle of double materiality. This enlarges the scope of reporting content, as the non-financial information is considered material if the impact of the company's operations on sustainability aspects is high ('inside-out perspective') and/or these sustainability issues affect the company's business development, performance and position ('outside-in perspective'). Implementing the CSRD and the EU taxonomy for classifying a company's sustainable economic activities challenges conventional corporate reporting and requires a change in internal reporting processes. While the objective of integrated reporting remains desirable for policymakers and stakeholders, some companies may find it difficult to embed this type of integrated thinking in their business in the short term. Still, to achieve the potential of integrated reporting, it is prudent for managers to proactively initiate the required internal transformation of the related processes, even if this will take some time to materialise.

This chapter concludes with five takeaways that could stimulate further discussion:

1. The expectations of stakeholders and shareholders regarding sustainability issues have changed significantly. If these sustainability expectations are understood as an opportunity for a more sustainable business development, the new reporting requirements can guide companies in this transition process.
2. The new reporting requirements oblige companies to report on their sustainability performance on an ongoing basis. The paradigm shift towards integrated thinking enables companies to align their strategic goals, decisions and performance indicators with their reporting obligations and thus present a more consistent picture to their stakeholders in the long term.
3. Integrated reporting offers the opportunity to explain in an understandable way, both internally and externally, how a company creates, preserves or destroys value, and to prevent information overload.
4. In light of the new reporting regulations (CSRD) and standards (ESRS) in the EU, the future role of the Integrated Reporting Framework remains unclear.
5. Reporting on non-financial performance indicators should be in line with regulatory frameworks and standards, but should be limited to material aspects to meet information requirements, remain manageable for companies (and auditors) and provide relevant, comparable and timely information to all stakeholders.

On the Road to Net Zero, strategy and reporting are the starting and ending points of operational business activities. Thus, the following three chapters will focus on related operational business areas that will enable the internal sustainability transformation. Chapter 5, *Creating Sustainable Products*, will further elaborate the paradigm shift in product development towards a circular economy.

References

Alniacik, U., Alniacik, E., & Genc, N. (2011). How corporate social responsibility information influences stakeholders' intentions. *Corporate Social Responsibility and Environmental Management, 18*(4), 234–245. https://doi.org/10.1002/csr.245

Bachmann, G. (2018). CSR-bezogene Vorstands- und Aufsichtsratspflichten und ihre Sanktionierungen. *Zeitschrift für Unternehmens und Gesellschaftsrecht, 47*(2–3), 231–261. https://doi.org/10.1515/zgr-2018-0010

Barrantes, E., Busch, F., Sckaer, M., & Zülch, H. (2022). Zukunft Geschäftsbericht—10 Thesen zum Wandel der Unternehmensberichterstattung.

KoR Zeitschrift für internationale und kapitalmarktorientierte Rechnungslegung, 22(2), 86–91.

Baumüller, J., & Schaffhauser-Linzatti, M.-M. (2018). In search of materiality for nonfinancial information—reporting requirements of the Directive 2014/95/EU. *NachhaltigkeitsManagementForum, 26*(1–4), 101–111. https://doi.org/10.1007/s00550-018-0473-z

Baumüller, J., Scheid, O., & Needham, S. (2021). Die Corporate Sustainability Reporting Directive als Schlüsselelement von Sustainable Finance: Zusammenhänge und Entwicklungsperspektiven. *Zeitschrift für Internationale Rechnungslegung, 16*(7–8), 337–343.

Bebbington, J., Larrinaga, C., & Moneva, J. (2008). Corporate social reporting and reputation risk management. *Accounting, Auditing & Accountability Journal, 21*(3), 337–361. https://doi.org/10.1108/09513570810863932

BMW AG. (Ed.) (2021). Our responsibility. Our future. BMW Group Report 2020. Retrieved June 19, 2023, from https://www.bmwgroup.com/content/dam/grpw/websites/bmwgroup_com/ir/downloads/en/2021/bericht/BMW-Group-Bericht-2020-EN.pdf

BMW AG. (Ed.) (2022a). BMW Group Report 2021. Retrieved March 02, 2023, from https://www.bmwgroup.com/content/dam/grpw/websites/bmwgroup_com/ir/downloads/en/2022/bericht/BMW-Group-Report-2021-en.pdf

BMW AG. (Ed.) (2022b). The future is electric, digital, and circular. BMW Group Report 2021. Retrieved June 19, 2023, from https://www.bmwgroup.com/content/dam/grpw/websites/bmwgroup_com/ir/downloads/en/2022/bericht/BMW-Group-Report-2021-en.pdf

BMW Group. (2014). Carbon Disclosure Project (CDP) zeichnet die BMW Group für Leistung und Transparenz zu Klimaschutzaktivitäten aus. Retrieved June 19, 2023, from https://www.press.bmwgroup.com/deutschland/article/detail/T0195046DE/carbon-disclosure-project-cdp-zeichnet-die-bmw-group-fuer-leistung-und-transparenz-zu-klimaschutzaktivitaeten-aus?language=de

BMW Group. (2021). BMW Group becomes first premium manufacturer to publish Integrated Group Report. Retrieved June 19, 2023, from https://www.press.bmwgroup.com/global/article/detail/T0327384EN/bmw-group-becomes-first-premium-manufacturer-to-publish-integrated-group-report?language=en

Böcking, H.-J., & Althoff, C. (2017). Paradigmenwechsel in der (Konzern-)Lageberichterstattung über nicht-monetäre Erfolgsfaktoren. *Der Konzern, 5*, 246–255.

Bonsón, E., & Bednárová, M. (2015). CSR reporting practices of Eurozone companies. *Revista de Contabilidad, 18*(2), 182–193. https://doi.org/10.1016/j.rcsar.2014.06.002

Brady, C., & Baraka, D. C. (2013). Integrated reporting. In S. O. Idowu, N. Capaldi, L. Zu, & A. D. Gupta (Eds.), *Encyclopedia of corporate social responsibility* (pp. 1460–1468). Springer. https://doi.org/10.1007/978-3-642-28036-8_341

Brundtland, G. H. (1987). Our common future—Call for action. *Environmental Conservation, 14*(4), 291–294. https://doi.org/10.1017/S0376892900016805

CDP. (2022). The A List 2022. Retrieved March 02, 2023, from https://www.cdp.net/en/companies/companies-scores

Chersan, I.-C. (2016). Corporate responsibility reporting according to Global Reporting Initiative: an international comparison. *Audit Financiar, 14*(136), 424–435. https://doi.org/10.20869/AUDITF/2016/136/424

Cormier, D., Magnan, M., & van Velthoven, B. (2005). Environmental disclosure quality in large German companies: Economic incentives, public pressures or institutional conditions? *European Accounting Review, 14*(1), 3–39. https://doi.org/10.1080/0963818042000339617

Council of the European Union. (2022). Council gives final green light to corporate sustainability reporting directive. Retrieved March 02, 2023, from Council of the European Union: https://www.consilium.europa.eu/en/press/press-releases/2022/11/28/council-gives-final-green-light-to-corporate-sustainability-reporting-directive/pdf

Deegan, C. (2002). Introduction: The legitimising effect of social and environmental disclosures—A theoretical foundation. *Accounting, Auditing & Accountability Journal, 15*(3), 282–311. https://doi.org/10.1108/09513570210435852

Diehl, T., & Knauß, R. (2018). Transparency through CSR reporting—the case of Stuttgart Airport. *Sustainability Management Forum, 26*(1–4), 113–121. https://doi.org/10.1007/s00550-018-0478-7

Druckmann, P., & Freis, J. (2010). Integrated reporting: The future of corporate reporting? In R. G. Eccles, B. Cheng, & D. Saltzmann (Eds.), *The landscape of integrated reporting. Reflections and next steps* (pp. 81–85). Harvard Business School. Retrieved June 01, 2023, from https://www.dvfa.de/fileadmin/_migrated/content_uploads/the_landscape_of_integrated_reporting.pdf

Du, S., Bhattacharya, C. B., & Sen, S. (2010). Maximizing business returns to corporate social responsibility (CSR): the role of CSR communication. *International Journal of Management Reviews, 12*(1), 8–19. https://doi.org/10.1111/j.1468-2370.2009.00276.x

Duchek, S. (2020). Organizational resilience: a capability-based conceptualization. *Business Research, 13*(1), 215–246. https://doi.org/10.1007/s40685-019-0085-7

Eccles, R. G., Cheng, B., & Saltzmann, D. (Eds.). (2010). *The landscape of integrated reporting: Reflections and next steps.* Harvard Business School. Retrieved June 01, 2023, from https://www.dvfa.de/fileadmin/_migrated/content_uploads/the_landscape_of_integrated_reporting.pdf

EFRAG. (2022). First set of draft ESRS. Retrieved March 02, 2023, from https://www.efrag.org/lab6

EFRAG. (2023). EFRAG welcomes the adoption of the delegated act on the first set of European Sustainability Reporting Standards (ESRS) by theEuropean Commission. Retrieved August 15, 2023, from https://www.efrag.org/News/

Public-439/EFRAG-welcomes-the-adoption-of-the-Delegated-Act-onthe-first-set-of-E?AspxAutoDetectCookieSupport=1

EFRAG PTF-ESRS. (Ed.) (2022). Draft European Sustainability Reporting Standards: ESRS 2 General disclosures. Retrieved June 10, 2023, from https://www.efrag.org/Assets/Download?assetUrl=%2Fsites%2Fwebpublishing%2FSiteAssets%2F07.%2520Draft%2520ESRS%25202%2520General%2520disclsoures%2520November%25202022.pdf

Eierle, B., & Kasischke, A. (2023). Finanzielle Berichterstattung über immaterielle Werte quo vadis?—Vorschläge und Würdigung des EFRAG-Diskussionspapiers. *KoR Zeitschrift für internationale und kapitalmarktorientierte Rechnungslegung, 23*(2), 69–77.

European Commission. (Ed.) (2021). FAQ: What is the EU Taxonomy and how will it work in practice? Retrieved March 02, 2023, from https://finance.ec.europa.eu/system/files/2021-04/sustainable-finance-taxonomy-faq_en.pdf

European Commission. (2022). Corporate sustainability reporting. Retrieved March 02, 2023, from https://finance.ec.europa.eu/capital-markets-union-and-financial-markets/company-reporting-and-auditing/company-reporting/corporate-sustainability-reporting_en

European Commission. (2023). Renewed sustainable finance strategy and implementation of the action plan on financing sustainable growth. Retrieved June 15, 2023, from https://finance.ec.europa.eu/publications/renewed-sustainable-finance-strategy-and-implementation-action-plan-financing-sustainable-growth_en

European Commission. (2023). The Commission adopts the European Sustainability Reporting Standards. Retrieved August 15, 2023, fromhttps://finance.ec.europa.eu/news/commission-adopts-european-sustainability-reporting-standards-2023-07-31_en

European Parliament. (2022). Sustainable economy: Parliament adopts new reporting rules for multinationals. Retrieved March 02, 2023, from European Parliament: https://www.europarl.europa.eu/news/en/press-room/20221107IPR49611/sustainable-economy-parliament-adopts-new-reporting-rules-for-multinationals

European Parliament and the Council of the European Union. (2014). Directive 2014/95/EU of the European Parliament and of the Council of 22 October 2014 amending Directive 2013/34/EU as regards disclosure of non-financial and diversity information by certain large undertakings and groups: Directive 2014/95/EU.

European Parliament and the Council of the European Union. (2020). Regulation (EU) 2020/852 of the European Parliament and of the Council of 18 June 2020 on the establishment of a framework to facilitate sustainable investment, and amending Regulation (EU) 2019/2088: Regulation (EU) 2022/852.

European Parliament and the Council of the European Union. (2022). Directive (EU) 2022/2464 of the European Parliament and of the Council of 14 December 2022 amending Regulation (EU) No 537/2014, Directive 2004/109/EC,

Directive 2006/43/EC and Directive 2013/34/EU, as regards corporate sustainability reporting: Directive (EU) 2022/2464.

Fischer, T. M., & Baumgartner, K. T. (2021). Understanding the reporting of intangibles from a business perspective. In B. Heidecke, M. C. Hübscher, R. Schmidtke, & M. Schmitt (Eds.), *Intangibles in the world of transfer pricing* (pp. 75–113). Springer. https://doi.org/10.1007/978-3-319-73332-6_6

Fitch Ratings. (2019). Fitch ratings launches ESG relevance scores to show impact of ESG on credit. Retrieved March 02, 2023, from https://www.fitchratings.com/research/corporate-finance/fitch-ratings-launches-esg-relevance-scores-to-show-impact-of-esg-on-credit-07-01-2019

Global Compact Network Germany & econsense. (Ed.) (2018). New momentum for reporting on sustainability: Study on implementation of the German CSR Directive Implementation Act. Retrieved March 02, 2023, from https://www.globalcompact.de/migrated_files/wAssets/docs/Reporting/NFE_Studie_Online_englisch_181015.pdf

GRI. (2022). Our mission and history. Retrieved March 02, 2023, from https://www.globalreporting.org/about-gri/mission-history/

Haller, A., & Fischer, T. M. (2023). Berichterstattung über Intangibles—Neue Impulse durch die CSRD. *KoR Zeitschrift für internationale und kapitalmarktorientierte Rechnungslegung, 23*(2), 78–87.

Hess, D. (2007). Social reporting and new governance regulation: The prospects of achieving corporate accountability through transparency. *Business Ethics Quarterly, 17*(03), 453–476. https://doi.org/10.5840/beq200717348

Hoffmann, E., Dietsche, C., & Hobelsberger, C. (2018). Between mandatory and voluntary: non-financial reporting by German companies. *Sustainability Management Forum, 26*(1–4), 47–63. https://doi.org/10.1007/s00550-018-0479-6

IFRS Foundation. (Ed.) (2022a). Integrated reporting framework. Retrieved March 02, 2023, from https://www.ifrs.org/issued-standards/ir-framework/

IFRS Foundation. (Ed.) (2022b). International <IR> Framework. Retrieved March 02, 2023, from http://www.integratedreporting.org/wp-content/uploads/2021/01/InternationalIntegratedReportingFramework.pdf

IIRC. (Ed.) (2019). Integrated thinking & strategy: State of play report. Retrieved March 02, 2023, from https://www.integratedreporting.org/wp-content/uploads/2020/01/Integrated-Thinking-and-Strategy-State-of-Play-Report_2020.pdf

Institut für ökologische Wirtschaftsforschung, & future e.V.—verantwortung unternehmen. (Eds.) (2022). CSR-Reporting in Deutschland 2021: Ergebnisse im Ranking der Nachhaltigkeitsberichte und Trends in der Berichterstattung von Großunternehmen und KMU. Retrieved March 02, 2023, from https://www.ranking-nachhaltigkeitsberichte.de/fileadmin/ranking/user_upload/2021/Ranking_Nachhaltigkeitsberichte_2021_Ergebnisbericht_1.pdf

IOSCO. (Ed.) (2019). Statement on disclosure of ESG matters by issuers. Retrieved March 02, 2023, from https://www.iosco.org/library/pubdocs/pdf/IOSCOPD619.pdf

Lock, I., & Seele, P. (2015). Analyzing sector-specific CSR reporting: Social and environmental disclosure to investors in the chemicals and banking and insurance industry. *Corporate Social Responsibility and Environmental Management, 22*(2), 113–128. https://doi.org/10.1002/csr.1338

Loew, T., & Braun, S. (2018). Mindestanforderungen und Obergrenzen für die Inhalte der nichtfinanziellen Erklärung: Interpretation der neuen HGB-regelungen zur nichtfinanziellen Berichterstattung aus Sicht der Lage- und der Nachhaltigkeitsberichterstattung. Empfehlungen an Unternehmen und Politik. Retrieved March 02, 2023, from https://www.4sustainability.de/wp-content/uploads/2021/06/Loew-Braun-Mindestanforderungen-Obergrenzen-nichtfinanzielle-Erklaerung-2018.pdf

Marquis, C., Toffel, M. W., & Zhou, Y. (2016). Scrutiny, norms, and selective disclosure: A global study of greenwashing. *Organization Science, 27*(2), 483–504. https://doi.org/10.1287/orsc.2015.1039

Moratis, L. (2018). Signalling responsibility? Applying signalling theory to the ISO 26000 standard for social responsibility. *Sustainability, 10*(11), 4172. https://doi.org/10.3390/su10114172

Navarrete-Oyarce, J., Gallegos, J. A., Moraga-Flores, H., & Gallizo, J. L. (2021). Integrated reporting as an academic research concept in the area of business. *Sustainability, 13*(14), 7741. https://doi.org/10.3390/su13147741

Rowbottom, N., & Locke, J. (2016). The emergence of <IR>. *Accounting and Business Research, 46*(1), 83–115. https://doi.org/10.1080/00014788.2015.1029867

Schmidt, M., & Strenger, C. (2019). NZG-Serie CSR: Die neuen nichtfinanziellen Berichtspflichten—Erfahrungen mit der Umsetzung aus Sicht institutioneller Investoren. *Neue Zeitschrift für Gesellschaftsrecht, 13*, 481–487.

Threlfall, R., King, A., Shulmann, J., & Bartels, W. (2020). The time has come: The KPMG Survey of Sustainability Reporting 2020. Retrieved March 02, 2023, from https://assets.kpmg.com/content/dam/kpmg/xx/pdf/2020/11/the-time-has-come.pdf

Union Investment. (Ed.) (2021). Nachhaltigkeit ist für die Mehrheit der Großanleger unverzichtbar—Ergebnisse der Markterhebung 2021 zum nachhaltigen Vermögensmanagement institutioneller Anleger in Deutschland. Retrieved March 02, 2023, from https://fmos.link/17274

Value Reporting Foundation. (2022a). Integrated thinking in action: A spotlight on the BMW Group. Retrieved March 02, 2023, from https://www.integratedreporting.org/wp-content/uploads/2022/01/VRF_Case_BMW-final-to-publish-28.1.2022.pdf

Value Reporting Foundation. (Ed.) (2022b). Integrated thinking principles: Value creation through organizational resilience. Retrieved March 02, 2023, from http://

www.integratedreporting.org/wp-content/uploads/2021/12/VRF_ITP-Main-120721.pdf

Velte, P. (2017). Die nichtfinanzielle Erklärung und die Diversity-Berichterstattung nach dem CSR-Richtlinie-Umsetzungsgesetz: Auf dem Weg zur "Sustainable Corporate Governance"? *StuB—Unternehmensteuern und Bilanzen, 19*(8), 293–298.

Velte, P., & Stawinoga, M. (2017). Integrated reporting: The current state of empirical research, limitations and future research implications. *Journal of Management Control, 28*(3), 275–320. https://doi.org/10.1007/s00187-016-0235-4

Zülch, H., Schneider, A., & Thun, T. (2023). CSRD-Incoming: Welche kapitalmarktorientierten Unternehmen sind für die neue Direktive bereit?: Top-Unternehmen, Implikationen und Empfehlungen. *KoR Zeitschrift für internationale und kapitalmarktorientierte Rechnungslegung, 23*(2), 99–101.

Open Access   This chapter is licensed under the terms of the Creative Commons Attribution 4.0 International License (http://creativecommons.org/licenses/by/4.0/), which permits use, sharing, adaptation, distribution and reproduction in any medium or format, as long as you give appropriate credit to the original author(s) and the source, provide a link to the Creative Commons license and indicate if changes were made.

The images or other third party material in this chapter are included in the chapter's Creative Commons license, unless indicated otherwise in a credit line to the material. If material is not included in the chapter's Creative Commons license and your intended use is not permitted by statutory regulation or exceeds the permitted use, you will need to obtain permission directly from the copyright holder.

5

Creating Sustainable Products
The Road to Circularity

Lena Ries, Sandro Wartzack, and Oliver Zipse

5.1 Introduction

Central to the circular economy (CE) is the shift from a linear cradle-to-grave system following a "take-make-use-dispose" approach toward a cradle-to-cradle system following a lifecycle approach (Lieder & Rashid, 2016). This implies the consideration of a product's entire life cycle along the value chain, including the extraction of raw materials, parts supply, manufacturing, distribution, and use, as well as end-of-life and waste management (Ellen MacArthur Foundation, 2013; Farooque et al., 2019). Thus, product design follows the principles of "design to redesign," where technical parts circulate in a closed system, and to "design out waste, pollutants, and emissions," where biological nutrients return to the biosphere (Murray et al., 2017). In the automotive industry, electric vehicles are discussed as a key technology for reducing greenhouse gas emissions (Li et al., 2019). In light of this transition from combustion vehicles toward the electrification of vehicles, the manufacturing phase and the downstream supply chain, rather than the use

---

L. Ries (✉)
FAU Erlangen-Nürnberg, Nuremberg, Germany
e-mail: lena.ries@fau.de

S. Wartzack
FAU Erlangen-Nürnberg, Erlangen, Germany

O. Zipse
BMW AG, Munich, Germany

phase, are decisive for the carbon footprint, as the manufacturing process of batteries for electric vehicles is highly energy intensive (Morfeldt et al., 2021). Thus, a lifecycle perspective is highly important for the creation of sustainability impact (Ries et al., 2023). Products must support circular strategies, such as maintenance, reuse, remanufacturing, or recycling, by intention, thereby emphasizing the importance of the designer's role (den Hollander et al., 2017).

On the Road to Net Zero outlined in this book, the actual implementation starts with product design once an integrated strategy and reporting scheme have been developed. Based on the principle of "what gets measured gets done" (see Chap. 1), the *Future of Corporate Disclosure* (see Chap. 4) enables companies to identify potential areas for improvement in their operations and products, including their sustainability performance. By integrating these insights into their product design processes, companies can create circular products that meet customer demands, corporate vision, and regulatory requirements.

The main objective of this chapter, *Creating Sustainable Products*, is to highlight the importance and implications of circular design on product and service development and to discuss the challenges faced by manufacturing companies in altering user behavior. The remainder of this chapter is organized into three sections. Section 5.2 starts with a short overview of the circularity concept. It then elaborates on three key implications of circularity for changing product design, service design, and user behavior. This is followed, in Sect. 5.3, by a conversation between Prof. Oliver Zipse, Chairman of the Board of Management of BMW AG, and Prof. Dr-Ing. Sandro Wartzack, Chair of Engineering Design at FAU Erlangen-Nürnberg. Both experts reflect on sustainable product appearance, globally varying customer expectations, and future advances in circular product design from a practitioner's perspective. Section 5.4 then gives an outlook on the future challenges of circular design before Sect. 5.5 concludes with a transitional link to the following Chap. 6 on *Transforming Value Chains for Sustainability*.

5.2 Pathways Toward Circular Design

A linear economy causes many of our current environmental problems, including natural resource depletion, biodiversity loss, and global warming (Rockström et al., 2009). For example, the extraction and processing of raw materials are responsible for 90% of global biodiversity loss and 50% of greenhouse gas emissions (International Resource Panel, 2019, p. 8). These environmental problems have presented the managers of manufacturing companies with immense difficulties. Climate change and resource scarcity in particular are placing manufacturing companies under increasing pressure to cope with new environmental regulations, resource price volatility, and supply chain risks (Gebhardt et al., 2022; Lieder & Rashid, 2016). One regulation proposed by the European Commission in 2020 is the Circular Economy Action Plan, which targets product design, the value retention of products and materials, and waste prevention (European Commission, 2022a). As a result, manufacturing companies now need to reconsider their conventional take-make-waste approaches (Geissdoerfer et al., 2017).

Taking a closer look at the automotive industry, 14% of global greenhouse emissions are attributed to transportation, and they keep rising (PWC, 2007). This is the result of two issues. First, current estimates indicate that the world fleet of vehicles will triple by 2050 compared to the base year 2000. Second, this fleet is aging, especially in developing countries, and is therefore not complying with stricter emission regulations (Mamalis et al., 2013). To tackle the environmental impact of the industry, companies need to adhere to increasing environmental regulations. For example, the new EU Battery Regulation, which is expected to come into force in 2023 (see European Parliament, 2023), for the first time, will set out rules concerning the entire life cycle of a product in terms of "production, recycling and repurposing" (European Commission, 2022b). In terms of production, new traction batteries for electric vehicles will have to be labeled to disclose their carbon footprints. In addition, value chain actors (except SMEs) will have to disclose that raw materials are responsibly sourced from a social and environmental point of view as part of a due diligence policy. Finally, the new digital battery passport, as well as stricter collection and recycling quotas, will foster reuse and recycling efforts (European Parliament, 2023).

As another example, the EU Commission has recently revealed its plans to revise the end-of-life vehicle (ELV) directive, which was initially enacted in 2000 (European Commission, 2023). This announcement marks a significant shift in the regulations that have governed ELVs for over 20 years. The proposed revision aims to bring about substantial changes in the way ELVs are collected, treated, and recycled, with the ultimate goal of aligning with the objectives of the European Green Deal. By encouraging the automotive industry to embrace a sustainable approach to car design and production, this initiative seeks to ensure consistency with the broader environmental goals of the European Union.

5.2.1 The CE Approach Offers a Paradigm Shift

In this context, scholars, politicians, and practitioners are promoting CE as a new paradigm that offers great potential (Mhatre et al., 2021). It offers new business opportunities to create value and employment while reducing material costs and price volatility (Kalmykova et al., 2018). Moreover, circular strategies can foster resource security (Stahel, 2016) and cut global greenhouse gas emissions by 63% by 2050 (Circle Economy, 2019). While the concept of CE was introduced by Pearce and Turner (1990), they used it to describe the relationship between the economy and nature, where nature provides inputs for production and serves as a sink for waste outputs (Geissdoerfer et al., 2017). This contrasts with the modern understanding of extending the life of resources (Blomsma & Brennan, 2017). The most prominent definition of CE currently in use has been provided by the Ellen McArthur Foundation (Kirchherr et al., 2017), which describes CE as "an industrial system that is restorative or regenerative by intention and design […]. It replaces the 'end-of-life' concept with restoration, shifts towards the use of renewable energy, eliminates the use of toxic chemicals, which impair reuse, and aims for the elimination of waste through the superior design of materials, products, systems, and, within this, business models" (Ellen MacArthur Foundation, 2013, p. 7). This definition highlights the importance of design to a CE in which the whole product life cycle, from design to end-of-life management, is considered (Farooque et al., 2019). Moreover, it shows how the understanding of CE is influenced by industrial ecology (Graedel & Allenby, 1995) and the cradle-to-cradle philosophy (McDonough & Braungart, 2003).

The cradle-to-cradle philosophy distinguishes two separable cycles: a biological cycle and a technical cycle. In the biological cycle, biodegradable materials provide nutrients for nature after use. In the technical cycle, the products and materials circulate in closed-loop industrial systems through processes such as reuse, repair, remanufacturing, and recycling. Consequently, waste no longer exists. The Ellen MacArthur Foundation visualizes this approach in the so-called butterfly diagram (Ellen MacArthur Foundation, 2019). Thus, in a closed-loop system, healthy and renewable resources are complemented with technical processes to retain product and material value over time. Three main principles guide the life cycle thinking of a CE: the first is to preserve and enhance natural capital, the second is to optimize resource yields, and the third is to foster system effectiveness (Ellen MacArthur Foundation, 2015).

5.2.2 Different Frameworks for CE Operationalization: Slowing, Closing, Narrowing, and R-Strategies

Manufacturing companies are considered pivotal for implementing a CE based on their potential to decouple value creation from resource use (Blomsma et al., 2019). As such, they can improve product use, extend product lifetime, and close materials flows, among others, through different circular strategies (Bocken et al., 2016; Potting et al., 2017). A variety of frameworks exist to operationalize CE principles for manufacturing companies. Bocken et al. (2016) describe three product design strategies, namely slowing resource loops (product durability and life-extending services), closing resource loops (recycling), and narrowing resource flows (resource efficiency), to manage material and product flows over time. Geissdoerfer et al. (2018) extend these strategies with intensifying resource loops (increased product use), and dematerialization of resource loops (substitution of product utility by service and software solutions).

Another approach to operationalizing a CE is to use the so-called R-strategies. While some authors distinguish between the three R's of reduce, reuse, and recycle (Ghisellini et al., 2016; Reike et al., 2018), others describe up to ten different R-strategies (Potting et al., 2017). While the former only addresses material flows, the latter includes a system perspective that addresses, for example, the rethinking of product use (Stumpf & Baumgartner, 2022). All varieties of the R-framework share a hierarchy that ranks the different R-strategies based on their value retention potential (Kirchherr et al., 2017; Reike et al., 2018). Strategies that aim at a useful application of materials are at the bottom of the hierarchy, while strategies that aim at extending product or component life are in the middle, and strategies that aim at intelligent production and use are at the top of the hierarchy (Stumpf & Baumgartner, 2022). An overview of the ten comprehensive R-strategies by Potting et al. (2017) is illustrated in Fig. 5.1, based on the visualization by Stumpf and Baumgartner (2022) and explained as follows: While *recovering* refers to energy, *recycling* describes the processing of materials to obtain the same or a lower quality of the material. Thus, these strategies address the material level and build the third cluster with the lowest priority for circularity. Strategies that extend product or component life comprise *repurposing*, which describes the use of products or components for a different function. Moreover, *refurbishing* (i.e., restoring and updating old products) and *remanufacturing* (i.e., using components of discarded products in a new product with the same function) are classified as

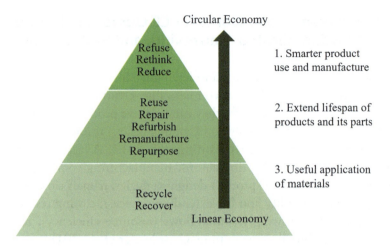

Fig. 5.1 R-strategies increasing circularity (own illustration based on Potting et al. (2017, p. 5) and Stumpf and Baumgartner (2022, p. 6))

life-extending strategies. Lastly, *repairing* defective products and *reusing* discarded products in good condition complement this cluster. The top cluster of smarter product use and manufacture comprises *reducing*, which implies an increase in efficiency in the manufacturing process or product use. Moreover, *rethinking*, which describes intensifying product use, and *refusing*, which implies making a product redundant by abandoning its function or by offering the same function with a radically different product, are based on business model innovation.

5.2.3 Three Implications for Design

Implementing the different R-strategies entails three implications for design: a change in product design, a change in service design, and a change in user behavior. Products need to be designed to embrace circular strategies (Bakker et al., 2014). However, circular products do not fulfill their potential if they end up in a drawer or landfill. That is why, in addition to changing the product design, manufacturers need to design new service offerings for reuse, repair, refurbishment, remanufacture, repurposing, and/or recycling (Revellio, 2022). These services must also be made attractive to the user if they are to actually be used (Amend et al., 2022), thereby emphasizing the role of the user and user behavior. Naturally, these three levels are interrelated, and trade-offs can occur among and within the three levels when aiming for circular design.

5.2.3.1 First Implication: A Change in Product Design

The Inertia Principle guides circular design following the hierarchy of the R-strategies: "Do not repair what is not broken, do not remanufacture something that can be repaired, do not recycle a product that can be remanufactured. […] [R]eplace or treat only the smallest possible part in order to maintain the existing economic value of the technical system" (Stahel, 2010, p. 195). This implies a product design for recirculation, endurance, and efficiency (Boyer et al., 2021).

At the product level, two key elements are important for the design dimension of *recirculation* (Boyer et al., 2021). First, increasing the fraction of a product that comes from used products (i.e., the input of recycled materials) (Linder et al., 2017) is also described as a design to reduce the embodied impact during production (Tecchio et al., 2017). This refers to the *recycling* strategy of the third cluster. Second, the fraction of recirculated outputs is relevant (i.e., how much of the product ends up being recirculated at the end of its functional life) (Boyer et al., 2021). In this context, the design for a technological cycle, the design for a biological cycle, and the design for disassembly and reassembly are relevant (Bocken et al., 2016). Thus, beyond the *recovering* and *recycling* strategies, the strategies of *refurbishing*, *remanufacturing*, and *repurposing* are important for a recirculated output, as they ensure reduced residual waste at the end of the functional life (Tecchio et al., 2017). An example of circularity in product design is BMW's "iVision Circular," a vehicle that is made as much as possible from secondary materials and is 100% recyclable at the end of life (BMW Group, 2021). Moreover, connectors and screws, instead of welds, are used where materials meet to facilitate easy disassembly. The potential of recirculation regarding carbon savings is promising. In China, BMW's joint venture works with local recycling companies to recover several materials, such as nickel, lithium, and cobalt, from spent high-voltage batteries and return them to the battery production cycle (BMW Group, 2022). According to BMW, this can save up to 70% of CO_2-emissions compared to using newly procured raw materials.

Another important product design dimension is *endurance*, which describes a product's ability to retain its value over time (Boyer et al., 2021). This requires, on the one hand, designing products for long life, including design for attachment and trust and design for reliability and durability. On the other hand, product design must ensure extended product use by including designs for maintenance and repair, designs for upgradability and adaptability, and designs for standardization and compatibility (Bocken et al., 2016; den Hollander et al. 2017). Design for modularity is also pivotal, as it allows the separation of modules of valuable parts that contain technology from those that do not (Krikke et al., 2004). Likewise, it facilitates the use of instruction manuals for self-repair (Amend et al., 2022). Thus, product endurance mainly relates to the circular strategies of *repairing*—designing for product life extension—and *rethinking*—designing for long-life products.

Designing for the *efficiency* of materials and resources during use stems from eco-design, and thus is not exclusive to the notion of CE (Tecchio et al., 2017). This type of design addresses the *reduce* strategy of the first cluster. An example of eco-efficient design in the automotive industry is lightweight design, which leads to a reduction in overall vehicle weight and increased fuel efficiency. The designers of BMW's first battery electric vehicle, the i3, used a carbon fiber-reinforced plastic body, which reduced weight by 50% compared to the use of steel in conventional car bodies (W. Zhang & Xu, 2022). Note that while this material increases efficiency during the use phase of the car due to the reduced weight, it hinders recycling at the end of life because of the material mix of plastics and carbon fiber. Moreover, it shifts emissions to the energy-intensive production of carbon fibers unless the production processes are powered by renewable energy, as in the case of BMW. This is an example of a trade-off in circular design on the product level and between life cycle stages.

In practice, the Circular Design Guide, developed collaboratively by the Ellen MacArthur Foundation and IDEO (Global Design & Innovation Company), provides methods and tools to help designers apply design thinking and circular design (Ellen MacArthur Foundation, 2017). To quantify product circularity, Linder et al. (2017) critically reviewed different metrics, such as the Material Circularity Indicator (MCI) developed by the Ellen MacArthur Foundation, the Cradle-to-Cradle (C2C) certification framework developed by the Cradle-to-Cradle Products Innovation Institute, or a circularity metric for products based on life cycle assessment (Scheepens et al., 2016). However, acquiring the necessary data for impact assessment can be difficult, and research must take on the challenges of developing accessible, unbiased, and easy-to-use tools (Boyer et al., 2021).

5.2.3.2 Second Implication: A Change in Service Design

Product-service systems (PSSs) are a type of business model that integrates tangible products and intangible services into a solution bundle to better satisfy customer needs (Mont, 2002). Based on their increasing servitization, PSSs are transitioning from a product focus toward providing services, product access, and performance (Tukker, 2004). Examples of this type of PSS are product-sharing systems, such as car sharing or appliance sharing (Bressanelli et al., 2018). This transition toward PSSs is relevant from an economic and environmental perspective (Tukker, 2015), as the design logic for PSS favors retaining product ownership to allow assessment of the total cost of ownership and designing for circularity (Tietze & Hansen, 2017). Therefore, PSSs are considered a means of dematerialization by paving the way for a more closed-loop, resource-efficient, and climate-friendly economy (Yang & Evans, 2019).

The use of PSSs has two implications for service design. First, the manufacturing companies need to design or form collaborations to offer additional services that can extend product lifetimes and close resource loops. These include services for maintenance, repair, upgrades, updates, take-back management, and waste handling processes (Lüdeke-Freund et al., 2019). These services need to be designed from a consumer-centric perspective, as discussed in the next section on user behavior. For example, circular services need to be included in a service contract (Amend et al., 2022), as users expect manufacturers to cover these costs (Mugge et al., 2005). Examples of these services in the automotive industry are the so-called re-factories of Renault in France and Spain, where used vehicles are refurbished, individual parts are remanufactured, traction batteries are repaired, and second-life applications are found for them (Groupe Renault, 2020).

The second requirement is that the manufacturing companies need to offer new PSS business models aimed at smarter product use (*refusing* and *rethinking*), such as product sharing (Bressanelli et al., 2018). In a sharing system, the service provider owns the product and therefore retains the responsibility for maintenance and repair, whereas different users can sequentially utilize the product and pay for this access (Tukker, 2004). These business models aim at intensified utilization, and respective metrics assess how often a product gets used (Boyer et al., 2021). However, their circularity impact depends on a change in user behavior (Tukker, 2015). For example, the potential of car-sharing business models to contribute to CO_2-emission reductions depends on the number of privately owned vehicles that are substituted for the car-sharing business model (Harris et al., 2021). The authors revealed that this is hardly the case at the moment due to rebound effects. For example, BMW found that the extent of environmental benefits depended on how services like car sharing were integrated into urban mobility ecosystems. Thus, the beneficial effects of on-demand mobility were very city-specific and depended on innovative and holistic transportation planning. This is why currently substituting private combustion engine cars for electric cars is the most CO_2-saving solution if the cars are charged with renewable energy. Changing user behavior is key to a positive circularity impact of service business models.

5.2.3.3 A Third Implication: A Change in User Behavior

The consumer's contribution to a CE has received little academic attention; however, as with products and services, user behavior must change from linear to circular (Selvefors et al., 2019). In a CE, the consumers have three roles (Shevchenko et al., 2023): First, they must select and buy a circular-oriented product or service rather than a conventional one. Second, they must not only use but also maintain and update the product. Lastly, they must discard the product through an appropriate channel for reuse, remanufacture, or recycling. Selvefors et al. (2019) describe these three phases from a user perspective, focusing on product exchange between users as obtaining the product (buying, trading, receiving products as gifts, leasing, subscribing, renting, borrowing, or co-using), using the product (utilizing, adjusting, repairing, repurposing, storing), and then resigning ownership of the product (gifting, trading, selling, returning a product to the provider, ending a lease or subscription contract, returning rented or borrowed products, or ending co-use). Based on this approach, the authors deduce user-centric design principles, including design for extended use, design for pre- and post-use, design for exchange, and design for multiple use cycles (Selvefors et al., 2019).

Therefore, research highlights the changing role of the consumer, who becomes a caretaker of the object in a CE (Rogers et al., 2021). This is similar to the notion of a pro-sumer (Kohtala, 2015) or pro-user (Stahel, 2019), who co-create products. In practice, however, evidence suggests that a tremendous gap exists between what people claim to do and how they actually behave. For example, 77% of European respondents said they undertake efforts to repair products, but 45% did not seek information on repairability (Parajuly et al., 2020). Therefore, designing for behavior change with the intent of influencing or promoting certain user behavior is pivotal for the implementation of a CE (Wastling et al., 2018). In this context, understanding the intrinsic (e.g., knowledge, motivation, habits, values) and extrinsic (e.g., norms, monetary incentives, infrastructural constraints) attributes that drive human behavior is important (Parajuly et al., 2020). By comparison, the car today is already one of the products that is kept alive for a long time by the established second-hand market, as well as by repairs.

To facilitate this behavioral change, two services are key. First, operational support is a service that supports the user in an efficient and durable product operation, such as training or performance monitoring (Kjaer et al., 2019). For example, well-designed repair manuals can help extend the product lifetime by aiding users in repairing rather than replacing a damaged product (Amend et al., 2022). Thus, operational support provides relevant knowledge and education on efficient product use. Second, behavioral support nudges users to act sustainably, thereby overcoming motivational challenges by setting and achieving goals (Ries et al., 2023). This can be achieved, for example, through positive feedback, gamification (e.g., repairability scores), monetary incentives, or a supporting community (Bovea et al., 2018; Valencia et al., 2015). Beyond fostering circular behavior, designing out adverse user behavior is equally important, as this can result in quicker wear and tear and decrease product longevity (Bressanelli et al., 2018). For example, in the case of a performance business model, customers might misuse products, thereby increasing maintenance costs, as these are covered by the provider (Reim et al., 2018). This link between pricing logic and user behavior emphasizes the need to understand how the pricing logic incentivizes certain behavior (Ries et al., 2023). For example, car-sharing pricing based on the minutes driven rather than on the distance driven is likely to incentivize fast, and therefore potentially unsafe, driving. Hence, the proper design of service contracts and pricing logic of service offers are pivotal for creating the desired circular behavior.

5.2.4 Implementation Challenges

For many companies, implementing circular strategies has not been easy (Lieder & Rashid, 2016), and this is especially the case with manufacturing companies (Lopes de Sousa Jabbour et al., 2018). In 2020, only 8.6% of the global economy was circularity oriented (Circle Economy, 2019, p. 8). One challenge is the required value network perspective, which requires enhancing relationships with supply chain actors, customers, and other service partners (Centobelli et al., 2020) to ensure the provision of additional services and PSS (Barreiro-Gen & Lozano, 2020). Compared to other industries, diverse services and PSSs are already associated with cars, from rental agencies and car repair workshops to used-car markets. Nevertheless, achieving full circularity requires additional collaboration. We will return to this idea in Sect. 5.3. Other barriers relate to governmental issues (e.g., the lack of standards), economic issues (e.g., the uncertainty regarding the profitability of circularity strategies), technological issues (e.g., design challenges in creating or maintaining durability), knowledge and skill issues (e.g., lack of skills), and management issues (e.g., lack of support from the top management) (Govindan & Hasanagic, 2018).

Currently, the focus of corporate efforts is centered on circular strategies involving reducing and recycling that combine environmental and economic benefits, particularly unilaterally, and it neglects the variety of circular strategies and an ecosystem approach (Barreiro-Gen & Lozano, 2020). For this reason, holistic implementation of circular strategies cannot be achieved solely through product design (Korhonen et al., 2018) and technological innovation (Suchek et al., 2021); it also requires stakeholder network (Evans et al., 2017) and learning (Bocken et al., 2018) perspectives. This also relates to the scope of the CE. While some perceive the CE as the operationalization for companies to implement sustainable development (Ghisellini et al., 2016; Murray et al., 2017), others perceive circularity as one archetype of sustainable business models (Allwood et al., 2012; Bocken et al., 2014). In a narrow sense, CE focuses on solutions that combine reduced environmental impact (resource efficiency and waste reduction) with increased economic value (customer value and growth). However, focusing only on these two dimensions—the ecology and the economy—fails to address all three dimensions of sustainability (Pieroni et al., 2019). Therefore, circular business models, being narrowly understood, might not always be sustainable. For CE to contribute to sustainable development, it must broaden its scope "from closed-loop recycling and short-term economic gains, towards a transformed economy that organises access to resources to maintain or enhance social well-being and environmental quality" (Velenturf & Purnell, 2021, p. 1453).

5.3 Expert Conversation on Sustainability in Product Development

Why Is It Important for BMW to Concentrate on Sustainability?

Zipse: Perhaps the most important ingredient in purchasing behavior is brand. We at BMW say that having a strong brand is very important—a brand with an innovative image, because the world very much links innovation with sustainability. We are convinced that most solutions for sustainable products come from innovation. Therefore, the impact of sustainability on brand image is the most important impact we have here.

Wartzack: What does this mean for product design?

Zipse: In addition to regulatory compliance, consumer behavior, and societal changes, it is about creating a brand image that remains attractive to current and future customers. We make sustainability one of the most important aspects of our product development because when it comes to products, you have to live up to what you say. You can talk a lot about what you want to achieve in the future or what your goals are. In product development, however, you have to put your words into action. People can experience your product, they can touch it, and of course they can drive it. People believe in your product strategy when they can see it.

What Is Your Customer Group?

Wartzack: In product development, we talk about Design for X, where X stands for recyclability, sustainability, use, transport, or production. Design for sustainability is very important, especially for the younger generation. We talk about Fridays for Future and CO_2-neutral production. However, the younger generation is not the typical BMW customer. What is your view on that?

Zipse: We have customers of all ages. They start at 25—these are really our new car buyers—and the average age is somewhere around 50. Across all age groups, sustainability becomes one of the most important factors in purchasing behavior. In other words, if your brand is not perceived as sustainable—especially a premium brand like BMW—you are out of the game. You are simply no longer attractive in this market. Sustainability is at the center of political movements around the world, and all stakeholders are realizing that innovation and sustainability are key.

Wartzack: Yes, I absolutely agree. It is important to make cars for all ages. Another thing that is changing: When we were young, it was important to own a car and to be free. The younger generation considers having a car very important when you need it—a connected car that is environmentally friendly.

What Are the Biggest Changes in Material Choice in Product Design, from a Conservative Focus on Cost and Functionality to a More Sustainability-Driven Approach?

Zipse: When we talk about product development, at the end of the day, it is about materials in the car. How do you see the use of materials in the car changing from, let's say, the old world, where cost and functionality were at the heart of product development, to a more sustainable approach?

Wartzack: There are many new materials, even natural materials like hemp, sisal, or flax. However, many design challenges arise with these new types of materials. For example, the maximum tensile stress for glass fiber is about 1000 megapascals, whereas for hemp fiber, it is 250 megapascals. Material engineers and product designers have to take this into account and design in a different way.

Zipse: I can relate to that. I think our engineering, innovation, and design departments face similar challenges with new natural materials. So, the question is: How do you overcome the design challenges associated with using natural materials?

Wartzack: With natural materials, you need reliable data. Why don't engineers like to design with wood, for example? Because it is very difficult to predict the behavior of wood, given its irregularities, such as knotholes. That is why design with natural materials is very difficult for the designer. But we have to differentiate. On the one hand, there are parts of the car in the main crash load path, where I would still use steel and aluminum, which are very recyclable. On the other hand, there are other parts, such as door systems or bulkheads, where biomaterials and biocomposites could be used. You can find concept cars in which the entire outer shell body is made of bio-composite materials. The key is to use the right material in the right place.

How Important Is Weight Reduction for Car Design?

Zipse: Weight has its ups and downs. When we were designing an electric car 10 years ago, we thought weight was the most important thing. Therefore, we used carbon fiber for the shell of the i3. We built the whole supply chain around carbon fiber, with a huge effort to make the car lighter. Making cars lighter is still a priority, but at the same time, other performance factors, such as aerodynamics, matter from a sustainability perspective. What is your view on weight reduction and lightweight materials?

Wartzack: Weight reduction through the use of high-quality lightweight materials is still an important factor, especially in the top-of-the-range segment, for reasons of driving dynamics. The BMW i3 impresses with its complete body made of carbon fiber-reinforced plastics, which was designed using very intensive dimensioning tools and computer-aided engineering tools. There is no doubt that designing with hybrid materials has huge advantages. However, their use requires new engineering skills and new recycling concepts. A car is a mixture of different materials, each put in the right place, depending on crash load paths and price.

How Can We Get More Natural Materials into Cars?

Wartzack: The interior of the i3 featured a lot of natural materials. This is very good for the user's perception. When you touch the surface, it feels warm. Are natural materials an important part of your future generation of cars?

Zipse: Natural materials are important, but we are also researching and developing future natural materials, such as synthetic leather, with materials that can eventually substitute crude oil. And we want to substitute natural leather in the end. Weight plays a role, too. By reducing weight, the car consumes less energy over its life cycle. However, we are increasingly seeing a secondary effect: if the car is lighter, less material is needed for its manufacture. If you look at the world today, it is all about resource efficiency. Today, humanity extracts around 100 billion tons of raw materials from the planet each year.

Wartzack: That seems to be the inconvenient truth.

Zipse: You can argue whether this is too much. Perhaps it is. It has already had the effect of steadily increasing the cost of extracting natural resources from the earth. So, in addition to weight reduction, a secondary approach is to use as little material as possible because raw materials are becoming more and more expensive. Look at palladium or rhodium these days. Of course, the COVID-19 pandemic was also a reason for the increase in raw material prices. But you are at risk: If you use too much material, your base cost will increase.

Wartzack: So what would be the right strategy to balance cost and weight reduction?

Zipse: We are very committed to reducing material and the base cost at the same time. We have implemented several methods to reduce weight. For example, we use bionic design and additive manufacturing technologies to build parts. Every kilogram of weight reduction in the car has another secondary effect: If the car is lighter, you can use smaller brakes or smaller battery packs to cover the same distance. That is why, after improving the car's aerodynamics, weight reduction is one of the most important areas of progress today. What do you think?

Wartzack: Yes, I absolutely agree. Sustainability means saving material and reducing weight. We can do a lot with the right dimensioning with the intensive use of dimensioning tools, for example. So, all in all, I think the product designers have to do their best to find a way to achieve these two goals.

What Would You Say Makes a Particular Material Sustainable?

Zipse: Do you think, when you choose materials, that a sustainable material choice has to look and feel sustainable? Is it a matter of haptics and quality? Is it enough that it is sustainably produced with a very small carbon footprint? Will customers pay extra money for sustainable features?

Wartzack: The appearance of a car is a very complex issue. On the one hand, the younger generations are striking on Fridays for Future, and society is demanding CO_2-neutral production. The world is waking up. On the other hand, everyone wants to buy an iPhone. User perception and attractiveness are very important for Apple products. Even older people buy iPhones despite the availability of mobile phones designed specifically for that age group. It's all about how much people love using your product. They like the design, the interaction, and the experience. Customers want to feel emotional about their products, and the integration of sustainable materials, the implementation of sustainable production, and sustainable supply chains are key arguments.

Zipse: You mentioned that the i3's interior materials feel warm. About 10 years ago, the interior had to be as cold as possible. Lots of chrome, lots of brushed metal, and so on—it was all over the place—black panels everywhere. That is changing. If you look at the iX or the new i7, it is designed more like a private lounge. There are almost no cold materials in your home anymore. Black leather is used less, and chrome is used less. Instead, we see warm, earthy materials. This also carries over to the interior of the car. It has a lot to do with the choice of materials. Electric cars are perceived as a space where you can withdraw from the outside world—especially because of the silent driving characteristics. People want to feel more at home. This has a big influence on the materials we choose.

Wartzack: And what do you think—how eco-friendly could a BMW look in the future?

How Eco-Friendly Will a BMW Look in the Future?

Zipse: We don't think it has to look like you are missing anything. It has to look like … You mentioned the iPhone. The iPhone is absolutely first class in terms of quality, and I think that will never go away. It is more the story you tell about how this product is made. It always has to look high quality. What you cannot do is neglect the quality of your product and claim it is sustainable. That will not work. There is no excuse for that. It is more a matter of what is perceived as aesthetically superior.

Wartzack: I absolutely agree with you about the quality aspect. But what is then perceived as superior by consumers today?

Zipse: Natural materials are perceived as aesthetic and progressive. You have to supply them at a very high level of quality. Then, the perception of quality comes naturally because the product is warm by its nature. But still, natural materials do not have to look natural. We are now on a level of interpretation that allows a lot more. You no longer see if it is based on natural or synthetic raw materials. It is all about design.

Wartzack: That is certainly the case with natural materials. They are sustainable and often recyclable or compostable, but what about synthetic materials? Especially in the interior design, components are made up of several layers, which limits their recyclability.

Zipse: Good point. Another important aspect is mono materials. My favorite example is a seat cover. It consists of a surface component, and material underneath, and the foam underneath is glued to another piece of foam. But they are two different materials. This makes them very difficult to recycle because they cannot be separated. What we are trying to do now with our next architecture is to use more mono materials. Mono materials are easy to recycle. These are things that we haven't thought about to this extent before. But the transition is quite easy: You have to start thinking from the recycling process, not just from the product design process. In the end, you may even find that you will have a better cost base.

How Have Recycling Approaches Evolved Over the Past Decades?

Wartzack: I remember the hype about recycling in the 1990s. I visited a pilot recycling plant in Munich and found it very impressive. Before that, I remember that some parts in the BMW dashboard were made out of polypropylene foam, PVC, and metal parts—a complete mixture of materials. So, a lot of Design for Recycling approaches and tools were developed in the 90s. How established are these approaches that have been developed since the 1990s in the BMW production environment today?

Zipse: In our private lives, we all know what a green dot is (in German "Grüner Punkt"—a sign for waste collection and recycling systems). Everyone knows that. What kind of material goes into which channel is regulated. Paper goes in this channel, mixed materials go in this channel, glass goes in that channel. The car industry is very big, but it has a highly diverse global regulatory landscape for the recycling process—the afterlife of the car. We expect some new regulations soon in the EU, where the Battery Regulation puts into place new objectives (e.g., for the recycled content). You have to think 10 years ahead about what will happen to our cars if a new policy is based on the upcoming revision of the End-of-Life vehicle directive. What happens to the car after the use phase? You can already start thinking about how to design your car if suddenly a policy is in place that requires that you recycle the car and extract all the raw materials. This immediately leads to the use of secondary materials. However, the quality of secondary materials today is not sufficient.

Wartzack: Why is that?

Zipse: Because they cannot be completely separated in the recycling process. If you could separate them the way we do with household waste in our daily lives, it would not be a problem. The recycled product would be as good as the product in the first cycle. Separating materials would be easy if you built separation into your product development strategy from the start. To do it in retrospect is extremely difficult. How can we increase the amount of secondary materials? You know the term "cradle to grave," but in the future, it should actually be "cradle to cradle." The car goes through its life cycle. At the end of the day, it is dismantled and recycled, and it is again part of a new car. A "cradle to cradle" system is the actual target we are aiming for. As a product design researcher, what are your suggestions on how to approach design for circularity?

Wartzack: The designer has to design in such a way that the materials can be easily recycled—by implementing detachable joints, for example. Transparency of information is also key. You need to know what material composition is behind some plastic labels because that can also be very confusing if they are not labeled correctly. The OEM should be in charge of recycling because they know their car best and can plan recycling strategies at the concept stage. The manufacturer knows best which components can be given a second life. BMW is already doing it for battery packs at its Leipzig plant, where used battery modules are used as stationary energy storage.

Zipse: Yes, we started using spent battery packs as stationary energy storage in 2017. So far, we are satisfied with the results of this pioneering project.

Wartzack: That's great to hear. A second life for cars is also a very sustainable approach and a good thing. If a car is used for 10–15 years in Germany today, it will then be used for another 10–20 years in Africa or elsewhere. This life cycle is quite common, as long as it is in line with existing recycling regulations. At first glance, this path does not necessarily seem to go against sustainability, but it does not bring secondary materials back into the cycle. I can imagine that ownership after the use phase is very difficult to control, and this also applies to effective recycling strategies when they are not mandated by legislation.

Zipse: I agree. Speaking of Africa and its resources, let us return to the use of natural materials. They have the best recycling properties, but at the same time, they are difficult to use because their properties are not so consistent. You mentioned the use of wood. What do you think about the product properties of natural materials and their recyclability?

How to Best Balance Between Natural Materials and Recyclability?

Wartzack: It is a big challenge. Using natural materials, such as natural fibers (e.g., hemp, flax, and sisal), wood, or leather, is a good thing, but it is challenging and makes life very complicated for the designer. That is clear. You need reliable data to make sound predictions about the mechanical properties of the materials. For example, performing your own tensile tests and simulations and conducting numerous cycles of validation, rather than blindly trusting the data sheets provided by the supplier, can be helpful. This is the basis for dimensioning products and even components. It is a lot of effort, but it pays off in the end. How do you deal with these challenges in BMW's product design department?

Zipse: We still need to learn more about how to use natural materials. One field of research and development (R&D) that needs to grow is simulation methodology—how do I simulate natural materials? That should even be a new field of research. As you said, it is worth the effort. Clean natural materials can go back into the natural cycle, while all other materials must be recycled for reuse, which is a task in itself.

Wartzack: It is actually an emerging field of research. With 100 billion tons of new raw materials to be extracted, the use of natural materials will not be the only answer, as their application in car design is limited.

Zipse: Well, you are right. And there are many other issues to bear in mind, such as the loss of biodiversity due to the additional land use needed to grow these amounts of materials if we use our current technologies. I learned a lot when I read Bill Gates's book because he put into perspective what we actually want to achieve by the year 2050. A lot of the technologies don't exist today; therefore, we have to conduct a lot of research to find the right technologies. None of the technologies that exist today are capable of solving our climate problem. I found that quite evident. Similarly, many aspects of car design still need to be rethought and require new technological developments.

Wartzack: To address these global challenges, we need more comprehensive approaches, such as Life Cycle Assessment (LCA). A lot can be done, but you need very accurate data throughout the entire value chain. Building up a sustainable process chain is quite tedious: you have to know where and how materials are extracted, how they are transported to the plant, and so on. Let me take the simple example of wine. What kind of wine would you prefer? For sustainability reasons, a wine from Australia or a wine from Italy? Most people would say, a wine from Italy, obviously. But imagine that the wine from Australia comes in large batches by ship. It could be that, at the end of the day, the wine has a better carbon footprint than the wine from Italy, which comes in small batches. Accuracy of data is key, so a lot of data analysis needs to be conducted to precisely measure and compare life cycles. The peak of recyclability and LCA approaches was in the 1990s, but today, we have completely new possibilities with AI. This is an emerging area of research.

Zipse: Wine is a very good example. Normally, one would assume that the Italian wine—or even better, a German wine—would have the best LCA because of the short transport distance. But this evidence is too simple. You learn from your mistakes: Ten years ago, we would have assumed that ride hailing was clearly good for the climate in cities.

Wartzack: And it is not?

Zipse: It actually turned out not to be the case, because first of all, with ride hailing, a lot of people switched from public transport to private transport, and consequently the number of kilometers driven increased. Traffic jams—our best example is San Francisco—have actually increased because of ride hailing. The assumption that this had only a positive environmental impact was incorrect. This is where we really have to understand and think all these things through to the end, to understand the whole life cycle effect. The life cycle effect will become a relevant decision factor in the future. Anything you do has a life cycle effect. Only by regulation will you see that life cycle effects become transparent and will be decisive.

Wartzack: There needs to be a regulatory effort to make lifecycle costs transparent. This becomes evident in the case of natural materials, which are sometimes more expensive than conventional materials because not all life cycle costs are captured.

Is It Worth Paying More for Natural Materials?

Zipse: We would if the entire life cycle effect was taken into account. At the end of the day, our cost base has to be in line with customer behavior. Customers today are extremely cost sensitive, even in the premium segment. This is not a one-size-fits-all answer, but if the market and consumers recognized a full life cycle effect, we would consider spending more money on it. We did this with carbon fiber, which was mainly produced using renewable energy sources in the i3. Of course, the carbon fiber structure was much more expensive than a normal steel structure. But we are open to making these bets for the future if we see evidence that it will have an overall life cycle effect that is acknowledged.

Is the Strategy Applied to the i3 Involving a Whole Structure Made from Carbon Fiber a Role Model for the Future?

Zipse: I think we have learned a lot about the use of carbon fiber. It is not necessarily scalable to very high volumes. Electromobility is now going to be a mass market segment, so an entire structure made of carbon fiber does not seem appropriate. However, you can see in the iX that the side frame is made from carbon fiber. We use it in certain structures where it makes sense, but there will not be another full carbon-fiber car body in the next few years. Product development is really one of our core competencies, and there are many exciting technological developments in the pipeline for the future.

5.4 The Future of Sustainable Product Development

As discussed in the Expert Discussion (Sect. 5.3), design for circularity is increasingly becoming the lynchpin of product development. While this paradigm shift offers many new opportunities for life cycle optimization, customer satisfaction, new business opportunities and recycling, it also presents challenges that need to be addressed. This section highlights two key aspects of the future of product development in the light of design for circularity.

5.4.1 Digital Technologies as Enablers of CE

Digital technologies, such as the Internet of Things (IoT), Big Data, Artificial Intelligence (AI), and Blockchain, can enable manufacturing companies to transition toward a CE (Chauhan et al., 2022). The support of digital technologies allows the collection of product lifetime information and the prediction of product condition and health status. This fosters the optimization and automation of business processes, thereby enabling different circular strategies (Alcayaga et al., 2019). Research suggests that the joint adoption of circular strategies and digital technologies increases firm performance (Lopes de Sousa Jabbour et al., 2022). For example, the digital product passport offers the possibility of storing static product information, such as material composition, disassembly instructions, and end-of-life handling, on a chip or sensor (Lopes de Sousa Jabbour et al., 2018). In addition, the passport can collect dynamic data, such as the product's history and alterations, during the product's life cycle (Hansen et al., 2020). Thus, a digital product passport enables the sharing of relevant data to facilitate different circular strategies, such as *recycling*.

In addition, regarding the second cluster of Fig. 5.1 (see Sect. 5.2), addressing lifetime extension, digital technologies can help to relocate used products and offer possibilities for the establishment of marketplaces in which former owners and second-hand buyers can trade products to enable the *reuse* strategy (Liu et al., 2022). Similarly, tracking and tracing product location and quality facilitates the harvesting of functioning modules or parts (Hansen et al., 2020). Regarding the *repair* strategy, the IoT and AI enable condition-based maintenance, which assesses the physical condition of a machine or product and deduces maintenance actions to prevent failure based on the derived insights (Ingemarsdotter et al., 2021). This has the potential to increase product performance, uptime, and lifespan (Alcayaga et al., 2019). Furthermore, algorithms and robotics can support efficient disassembly, depending on the quality of the product and its parts, for *refurbishing* or *remanufacturing* (Hansen et al., 2020; Kerin & Pham, 2020). Lastly, just as in the case of the reuse strategy, marketplaces based on platform technologies can enable *repurposing* by transforming wastes or byproducts created in one industry into production inputs for other industries (Liu et al., 2022).

Additionally, to *reduce* the environmental impact of products at the product development stage, designers can use simulation methods. Similarly, modeling tools can help to better understand the sustainability impacts of decisions made in product design (e.g., the choice of the material composition for the product) by testing multiple interactions between the environmental, social, and economic dimensions (Jaghbeer et al., 2017). In addition, other technologies, such as the digital twin, offer the opportunity to predict and control carbon emissions by optimizing the manufacturing process (C. Zhang & Ji, 2019). Regarding *rethinking* and *refusing* strategies, offering new services and altering user behavior is key. Digital technologies can help to change user behavior, such as supporting an efficient and sustainable use to foster longevity, by monitoring and incentivizing user behavior (Bressanelli et al., 2018) and enabling operational and behavioral support (Ries et al., 2023).

5.4.2 Better Together: The Need for Broadening Perspectives

As mentioned earlier, extending the implementation of circular strategies cannot rely only on product design (Korhonen et al., 2018) and technological innovation (Suchek et al., 2021). Instead, a shift is needed in doing business to expand impact assessment and address the whole product life cycle, including end-of-life and social aspects (Farooque et al., 2019). The organizational boundaries also need expansion to embrace stakeholder collaboration along the value chain (Evans et al., 2017).

Evaluation of the sustainability impact of circular business models requires analysis of a variety of effects and trade-offs between and within lifecycle stages early on. First, rebound effects can cause detrimental sustainability effects (Kjaer et al., 2019). For example, circular strategies can lead to lower prices, less time consumption, or more accessible services that, in turn, increase demand, ultimately leading to an increase in resource consumption, waste, and emissions (Castro et al., 2022). Negative consumption-based shifts between life cycle stages (Kjaer et al., 2016) or trade-offs within one or between different design elements (Ries et al., 2023) can also occur. For example, energy consumption might increase as maintenance processes are optimized based on digital technologies (Halstenberg et al., 2019). Lastly, rebalancing effects might arise. These describe, for example, the activity of relocating bicycles with the help of vehicles and staff to compensate for asymmetric use patterns in product sharing (Bonilla-Alicea et al., 2020). Thus, a thorough understanding of the life cycle is necessary, complemented with an understanding of underlying assumptions regarding behavior, for any analysis of the sustainability effects produced by circular business models (Niero et al., 2021).

This understanding must consider both the environmental impact and the social impact, thereby extending the scope of CE to embrace all sustainability dimensions. While sufficient indicators are available for social life cycle analysis, most studies have focused on indicators related to health and safety at the workplace of focal companies while neglecting value chain actors and consumers (Kühnen & Hahn, 2017). Digital technologies can help to consider the social dimension of sustainability in these assessments. For example, a combination of digital technologies can help to analyze product stewardship (i.e., health and safety effects on the user) (Ries et al., 2023). Examples are injury prevention (Moreno et al., 2017), breakdown avoidance (Lim et al., 2018), safe driving (Haftor & Climent, 2021), and healthy living (Valencia et al., 2015). Blockchain technology can further increase the willingness of value chain actors to share confidential social data needed for these assessments (Rusch et al., 2022).

This aspect relates to the expansion of boundaries embracing collaboration. While the integration of stakeholders and coordination among partners in the business ecosystem become a crucial skill in the transition toward a CE (Santa-Maria et al., 2022), the development of ecosystems and value co-creation within them, based on connectivity and interactivity, poses a challenge for many companies (A. Q. Li et al., 2020). A business ecosystem for circularity comprises a set of actors that include producers, suppliers, service providers, end users, collectors, disassemblers, recyclers, policymakers, and members of civil society organizations who contribute to a collective outcome (Konietzko et al., 2020). Building this ecosystem requires that manufacturers engage with regulatory bodies to develop better circular strategies (Awan et al., 2021), but they must also interact with collectors, dismantlers, and recyclers to increase efficiency and reduce recycling costs (Parida et al., 2019).

Product collectors, dismantling companies, and recyclers are crucial actors in a circular supply chain (Lüdeke-Freund et al., 2019). Feedback and circular involvement from the end-of-life phase to the product design phase of the OEM are important for comprehensive leveraging of circular potentials (Hansen & Revellio, 2020); however, the information flow usually ends with the user (Blömeke et al., 2020). Manufacturers, users, reverse logistic providers, dismantling companies, and recyclers can overcome these deficits by forming connections through smart devices and digital platforms, thereby increasing collection, dismantling, and recycling efficiency (Liu et al., 2022). By facilitating collaboration and automation, digital technologies can improve product disassembly and recycling and contribute to economic feasibility (Blömeke et al., 2020). One example of this type of a new data ecosystem in the automotive industry is Catena-X, where different value chain actors are currently building a platform to enable the crucial information exchange on product history and the state of health of the vehicle and its components (Mügge et al., 2023). Implementing new technology advancements for an optimized data exchange might have the potential to support the formation and expansion of circular business ecosystems.

Establishing and tightening relationships can create a common understanding among different stakeholders and foster circular strategy implementation (Schöggl et al., 2020).

5.5 Conclusion

The CE addresses current challenges of resource scarcity, global warming, and economic volatility. To operationalize this abstract concept, the ten R-strategies of refusing, rethinking, reducing (smarter product use and manufacture), reusing, repairing, refurbishing, remanufacturing, repurposing (extend the lifespan of products and their parts), recycling, and recovering (useful application of materials) are widely recognized. Their implementation has implications for design.

How does one approach design for circularity? We want to highlight five takeaways from this chapter that invite further discussion:

1. Circular design requires a change in product design. To fulfill circularity principles, innovators need to incorporate recycled materials and consider the future circulation of the product, its parts, and its materials. The use of recycled materials requires that the quality of the recyclate matches the given requirements of the product and that the material be available in sufficient quantity. Designing for endurance and efficiency further complements the aspect of recirculation. Thus, recirculation, endurance, and efficiency serve as guides for product design.
2. Circular design requires a change in service design. Additional services, such as take-back, enable circularity. Alternatively, a shift in ownership from the producer to the provider and toward access and performance business models can implement circular strategies. Thus, product design must go hand-in-hand with service design.
3. Circular design requires a change in user behavior. The CE emphasizes the role of the consumer as a crucial actor who takes care of the product and returns it at a given time in the life cycle. User behavior is a key determinant in assessing any sustainable impact. Thus, circular design needs to be especially user-centric.
4. Circular design requires a social dimension. So far, industry and research have focused on the environmental–economic nexus within the CE concept. However, extending the scope to embrace the social dimension is the key to sustainable development.
5. Circular design requires collaboration along the value chain. To implement circular strategies, manufacturing companies need to extend organizational boundaries and establish business ecosystems that include, for example, suppliers, service providers, logistic providers, customers,

dismantlers, and end-of-life vehicle recyclers. A new mindset is necessary that fosters openness to collaboration and lifecycle thinking.

Circular design, such as the use of recyclable or renewable materials and the development of new services to close resource loops, starts with product and service innovation, but it needs to embrace many different functions within a company and a variety of actors across organizational boundaries spanning an automotive ecosystem. This results from the need to shift organizational thinking from a product to a PSS, from a production to a lifecycle, and from an individual to a collaborative approach. The next stage (Chap. 6) on the *Road to Net Zero* focuses on *Transforming Value Chains for Sustainability*.

References

Alcayaga, A., Wiener, M., & Hansen, E. (2019). Towards a framework of smart-circular systems: An integrative literature review. *Journal of Cleaner Production, 221*, 622–634. https://doi.org/10.1016/j.jclepro.2019.02.085

Allwood, J. M., Cullen, J. M., & Carruth, M. A. (2012). *Sustainable materials: With both eyes open*. UIT Cambridge.

Amend, C., Revellio, F., Tenner, I., & Schaltegger, S. (2022). The potential of modular product design on repair behavior and user experience—Evidence from the smartphone industry. *Journal of Cleaner Production, 367*, 132770. https://doi.org/10.1016/j.jclepro.2022.132770

Awan, U., Sroufe, R., & Shahbaz, M. (2021). Industry 4.0 and the circular economy: A literature review and recommendations for future research. *Business Strategy and the Environment, 30*(4), 2038–2060. https://doi.org/10.1002/bse.2731

Bakker, C., Wang, F., Huisman, J., & den Hollander, M. (2014). Products that go round: exploring product life extension through design. *Journal of Cleaner Production, 69*, 10–16. https://doi.org/10.1016/j.jclepro.2014.01.028

Barreiro-Gen, M., & Lozano, R. (2020). How circular is the circular economy? Analysing the implementation of circular economy in organisations. *Business Strategy and the Environment, 29*(8), 3484–3494. https://doi.org/10.1002/bse.2590

Blömeke, S., Mennenga, M., Herrmann, C., Kintscher, L., Bikker, G., Lawrenz, S., Sharma, P., Rausch, A., Nippraschk, M., Goldmann, D., Poschmann, H., Brüggemann, H., Scheller, C., & Spengler, T. (2020). Recycling 4.0. In R. Chitchyan, D. Schien, A. Moreira, & B. Combemale (Eds.), *Proceedings of the 7th international conference on ICT for sustainability* (pp. 66–76). ACM. Retrieved on June 19, 2023, from https://doi.org/10.1145/3401335.3401666

Blomsma, F., & Brennan, G. (2017). The emergence of circular economy: A new framing around prolonging resource productivity. *Journal of Industrial Ecology, 21*(3), 603–614. https://doi.org/10.1111/jiec.12603

Blomsma, F., Pieroni, M., Kravchenko, M., Pigosso, D. C., Hildenbrand, J., Kristinsdottir, A. R., Kristoffersen, E., Shahbazi, S., Nielsen, K. D., Jönbrink, A.-K., Li, J., Wiik, C., & McAloone, T. C. (2019). Developing a circular strategies framework for manufacturing companies to support circular economy-oriented innovation. *Journal of Cleaner Production, 241*, 118271. https://doi.org/10.1016/j.jclepro.2019.118271

BMW Group. (2021, September 6). Der BMW i vision circular [Press release]. München. Retrieved June 19, 2023, from https://www.press.bmwgroup.com/deutschland/article/detail/T0341253DE/der-bmw-i-vision-circular?language=de

BMW Group. (2022, May 25). BMW Group etabliert geschlossenen Recycling-Kreislauf für Hochvoltbatterien in China [Press release]. München. Retrieved June 19, 2023, from https://www.press.bmwgroup.com/deutschland/article/detail/T0393733DE/bmw-group-etabliert-geschlossenen-recycling-kreislauf-fuer-hochvoltbatterien-in-china

Bocken, N., de Pauw, I., Bakker, C., & van der Grinten, B. (2016). Product design and business model strategies for a circular economy. *Journal of Industrial and Production Engineering, 33*(5), 308–320. https://doi.org/10.1080/21681015.2016.1172124

Bocken, N., Short, S. W., Rana, P., & Evans, S. (2014). A literature and practice review to develop sustainable business model archetypes. *Journal of Cleaner Production, 65*, 42–56. https://doi.org/10.1016/j.jclepro.2013.11.039

Bocken, N., Schuit, C., & Kraaijenhagen, C. (2018). Experimenting with a circular business model: Lessons from eight cases. *Environmental Innovation and Societal Transitions, 28*, 79–95. https://doi.org/10.1016/j.eist.2018.02.001

Bonilla-Alicea, R. J., Watson, B. C., Shen, Z., Tamayo, L., & Telenko, C. (2020). Life cycle assessment to quantify the impact of technology improvements in bike-sharing systems. *Journal of Industrial Ecology, 24*(1), 138–148. https://doi.org/10.1111/jiec.12860

Bovea, M. D., Quemades-Beltrán, P., Pérez-Belis, V., Juan, P., Braulio-Gonzalo, M., & Ibáñez-Forés, V. (2018). Options for labelling circular products: Icon design and consumer preferences. *Journal of Cleaner Production, 202*, 1253–1263. https://doi.org/10.1016/j.jclepro.2018.08.180

Boyer, R. H. W., Mellquist, A.-C., Williander, M., Fallahi, S., Nyström, T., Linder, M., Algurén, P., Vanacore, E., Hunka, A. D., Rex, E., & Whalen, K. A. (2021). Three-dimensional product circularity. *Journal of Industrial Ecology, 25*(4), 824–833. https://doi.org/10.1111/jiec.13109

Bressanelli, G., Adrodegari, F., Perona, M., & Saccani, N. (2018). Exploring how usage-focused business models enable circular economy through digital technologies. *Sustainability, 10*(3), 639. https://doi.org/10.3390/su10030639

Castro, C. G., Trevisan, A. H., Pigosso, D. C., & Mascarenhas, J. (2022). The rebound effect of circular economy: Definitions, mechanisms and a research agenda. *Journal of Cleaner Production, 345*, 131136. https://doi.org/10.1016/j.jclepro.2022.131136

Centobelli, P., Cerchione, R., Chiaroni, D., Del Vecchio, P., & Urbinati, A. (2020). Designing business models in circular economy: A systematic literature review and research agenda. *Business Strategy and the Environment, 29*(4), 1734–1749. https://doi.org/10.1002/bse.2466

Chauhan, C., Parida, V., & Dhir, A. (2022). Linking circular economy and digitalisation technologies: A systematic literature review of past achievements and future promises. *Technological Forecasting and Social Change, 177*, 121508. https://doi.org/10.1016/j.techfore.2022.121508

Circle Economy. (2019). The circularity gap report 2019. Retrieved June 19, 2023, from https://www.circle-economy.com/resources/the-circularity-gap-report-2019

den Hollander, M. C., Bakker, C. A., & Hultink, E. J. (2017). Product design in a circular economy: development of a typology of key concepts and terms. *Journal of Industrial Ecology, 21*(3), 517–525. https://doi.org/10.1111/jiec.12610

Ellen MacArthur Foundation. (2013). Towards the circular economy: economic and business rationale for an accelerated transition. Retrieved June 19, 2023, from https://ellenmacarthurfoundation.org/towards-the-circular-economy-vol-1-an-economic-and-business-rationale-for-an

Ellen MacArthur Foundation. (2015). Growth within: a circular economy vision for a competitive Europe. Retrieved June 19, 2023, from https://ellenmacarthurfoundation.org/growth-within-a-circular-economy-vision-for-a-competitive-europe

Ellen MacArthur Foundation. (2017). The circular design guide. Retrieved June 19, 2023, from https://www.circulardesignguide.com/

Ellen MacArthur Foundation. (2019). The butterfly diagram: visualizing the circular economy. Retrieved June 19, 2023, from https://ellenmacarthurfoundation.org/circular-economy-diagram

European Commission. (2022a). Circular economy action plan. Retrieved June 19, 2023, from https://environment.ec.europa.eu/strategy/circular-economy-action-plan_en

European Commission. (2022b, December 9). Green Deal: EU agrees new law on more sustainable and circular batteries to support EU's energy transition and competitive industry [Press release]. Retrieved June 19, 2023, from https://ec.europa.eu/commission/presscorner/detail/en/ip_22_7588

European Commission. (Ed.) (2023). End-of-life vehicles—revision of EU rules. Retrieved June 19, 2023, from https://ec.europa.eu/info/law/better-regulation/have-your-say/initiatives/12633-End-of-life-vehicles-revision-of-EU-rules_en

European Parliament. (2023). Making batteries more sustainable and better-performing [Press release]. Retrieved June 19, 2023, from https://www.europarl.europa.eu/news/en/agenda/briefing/2023-06-12/4/making-batteries-more-sustainable-more-durable-and-better-performing

Evans, S., Vladimirova, D., Holgado, M., van Fossen, K., Yang, M., Silva, E. A., & Barlow, C. Y. (2017). Business model innovation for sustainability: towards a unified perspective for creation of sustainable business models. *Business Strategy and the Environment, 26*(5), 597–608. https://doi.org/10.1002/bse.1939

Farooque, M., Zhang, A., Thürer, M., Qu, T., & Huisingh, D. (2019). Circular supply chain management: A definition and structured literature review. *Journal of Cleaner Production, 228*, 882–900. https://doi.org/10.1016/j.jclepro.2019.04.303

Gebhardt, M., Spieske, A., & Birkel, H. (2022). The future of the circular economy and its effect on supply chain dependencies: Empirical evidence from a Delphi study. *Transportation Research Part E: Logistics and Transportation Review, 157*, 102570. https://doi.org/10.1016/j.tre.2021.102570

Geissdoerfer, M., Morioka, S. N., de Carvalho, M. M., & Evans, S. (2018). Business models and supply chains for the circular economy. *Journal of Cleaner Production, 190*, 712–721. https://doi.org/10.1016/j.jclepro.2018.04.159

Geissdoerfer, M., Savaget, P., Bocken, N., & Hultink, E. J. (2017). The circular economy—A new sustainability paradigm? *Journal of Cleaner Production, 143*, 757–768. https://doi.org/10.1016/j.jclepro.2016.12.048

Ghisellini, P., Cialani, C., & Ulgiati, S. (2016). A review on circular economy: the expected transition to a balanced interplay of environmental and economic systems. *Journal of Cleaner Production, 114*, 11–32. https://doi.org/10.1016/j.jclepro.2015.09.007

Govindan, K., & Hasanagic, M. (2018). A systematic review on drivers, barriers, and practices towards circular economy: a supply chain perspective. *International Journal of Production Research, 56*(1–2), 278–311. https://doi.org/10.1080/00207543.2017.1402141

Graedel, T., & Allenby, B. R. (1995). *Industrial ecology*. Prentice Hall.

Groupe Renault. (2020, November). Refactory [Press release]. Retrieved June 19, 2023, from https://www.press.renault.co.uk/assets/documents/original/18432-REFACTORYGROUPERENAULTFLINSPLANTPRESSKIT251120.pdf

Haftor, D. M., & Climent, R. C. (2021). CO2 reduction through digital transformation in long-haul transportation: Institutional entrepreneurship to unlock product-service system innovation. *Industrial Marketing Management, 94*, 115–127. https://doi.org/10.1016/j.indmarman.2020.08.022

Halstenberg, F. A., Lindow, K., & Stark, R. (2019). Leveraging circular economy through a methodology for smart service systems engineering. *Sustainability, 11*(13), 3517. https://doi.org/10.3390/su11133517

Hansen, E., & Revellio, F. (2020). Circular value creation architectures: Make, ally, buy, or laissez-faire. *Journal of Industrial Ecology, 24*(6), 1250–1273. https://doi.org/10.1111/jiec.13016

Hansen, E., Wiedemann, P., Fichter, K., Lüdeke-Freund, F., Jaeger-Erben, M., Schomerus, T., Alcayaga, A., Blomsma, F., Tischner, U., Ahle, U., Büchle, D., Denker, A., Fiolka, K., Fröhling, M., Häge, A., Hoffmann, V., Kohl, H., Nitz, T., Schiller, C., … Kadner, S. (2020). Circular business models: Overcoming barriers, unleashing potentials. acatech/Circular Economy Initiative Deutschland/ SYSTEMIQ.

Harris, S., Mata, É., Plepys, A., & Katzeff, C. (2021). Sharing is daring, but is it sustainable? An assessment of sharing cars, electric tools and offices in Sweden. *Resources, Conservation and Recycling, 170*, 105583. https://doi.org/10.1016/j.resconrec.2021.105583

Ingemarsdotter, E., Kambanou, M. L., Jamsin, E., Sakao, T., & Balkenende, R. (2021). Challenges and solutions in condition-based maintenance implementation—A multiple case study. *Journal of Cleaner Production, 296*, 126420. https://doi.org/10.1016/j.jclepro.2021.126420

International Resource Panel. (2019). Global resources outlook 2019: Natural resources for the future we want. Retrieved June 19, 2023 from https://www.resourcepanel.org/reports/global-resources-outlook

Jaghbeer, Y., Hallstedt, S. I., Larsson, T., & Wall, J. (2017). Exploration of simulation-driven support tools for sustainable product development. *Procedia CIRP, 64*, 271–276. https://doi.org/10.1016/j.procir.2017.03.069

Kalmykova, Y., Sadagopan, M., & Rosado, L. (2018). Circular economy—From review of theories and practices to development of implementation tools. *Resources, Conservation and Recycling, 135*, 190–201. https://doi.org/10.1016/j.resconrec.2017.10.034

Kerin, M., & Pham, D. T. (2020). Smart remanufacturing: a review and research framework. *Journal of Manufacturing Technology Management, 31*(6), 1205–1235. https://doi.org/10.1108/JMTM-06-2019-0205

Kirchherr, J., Reike, D., & Hekkert, M. (2017). Conceptualizing the circular economy: An analysis of 114 definitions. *Resources, Conservation and Recycling, 127*, 221–232. https://doi.org/10.1016/j.resconrec.2017.09.005

Kjaer, L. L., Pagoropoulos, A., Schmidt, J. H., & McAloone, T. C. (2016). Challenges when evaluating product/service-systems through life cycle assessment. *Journal of Cleaner Production, 120*, 95–104. https://doi.org/10.1016/j.jclepro.2016.01.048

Kjaer, L. L., Pigosso, D. C. A., Niero, M., Bech, N. M., & McAloone, T. C. (2019). Product/service-systems for a circular economy: the route to decoupling economic growth from resource consumption? *Journal of Industrial Ecology, 23*(1), 22–35. https://doi.org/10.1111/jiec.12747

Kohtala, C. (2015). Addressing sustainability in research on distributed production: an integrated literature review. *Journal of Cleaner Production, 106*, 654–668. https://doi.org/10.1016/j.jclepro.2014.09.039

Konietzko, J., Bocken, N., & Hultink, E. J. (2020). Circular ecosystem innovation: An initial set of principles. *Journal of Cleaner Production, 253*, 119942. https://doi.org/10.1016/j.jclepro.2019.119942

Korhonen, J., Nuur, C., Feldmann, A., & Birkie, S. E. (2018). Circular economy as an essentially contested concept. *Journal of Cleaner Production, 175*, 544–552. https://doi.org/10.1016/j.jclepro.2017.12.111

Krikke, H., Le Blanc, I., & van de Velde, S. (2004). Product modularity and the design of closed-loop supply chains. *California Management Review, 46*(2), 23–39. https://doi.org/10.2307/41166208

Kühnen, M., & Hahn, R. (2017). Indicators in social life cycle assessment: A review of frameworks, theories, and empirical experience. *Journal of Industrial Ecology, 21*(6), 1547–1565. https://doi.org/10.1111/jiec.12663

Li, A. Q., Rich, N., Found, P., Kumar, M., & Brown, S. (2020). Exploring product–service systems in the digital era: a socio-technical systems perspective. *The TQM Journal, 32*(4), 897–913. https://doi.org/10.1108/TQM-11-2019-0272

Li, Z., Khajepour, A., & Song, J. (2019). A comprehensive review of the key technologies for pure electric vehicles. *Energy, 182*, 824–839. https://doi.org/10.1016/j.energy.2019.06.077

Lieder, M., & Rashid, A. (2016). Towards circular economy implementation: a comprehensive review in context of manufacturing industry. *Journal of Cleaner Production, 115*, 36–51. https://doi.org/10.1016/j.jclepro.2015.12.042

Lim, C.-H., Kim, M.-J., Heo, J.-Y., & Kim, K.-J. (2018). Design of informatics-based services in manufacturing industries: case studies using large vehicle-related databases. *Journal of Intelligent Manufacturing, 29*(3), 497–508. https://doi.org/10.1007/s10845-015-1123-8

Linder, M., Sarasini, S., & van Loon, P. (2017). A Metric for Quantifying Product-Level Circularity. *Journal of Industrial Ecology, 21*(3), 545–558. https://doi.org/10.1111/jiec.12552

Liu, Q., Trevisan, A. H., Yang, M., & Mascarenhas, J. (2022). A framework of digital technologies for the circular economy: Digital functions and mechanisms. *Business Strategy and the Environment, 31*(5), 2171–2192. https://doi.org/10.1002/bse.3015

Lopes de Sousa Jabbour, A. B., Chiappetta Jabbour, C. J., Choi, T.-M., & Latan, H. (2022). 'Better together': Evidence on the joint adoption of circular economy and industry 4.0 technologies. *International Journal of Production Economics, 252*, 108581. https://doi.org/10.1016/j.ijpe.2022.108581

Lopes de Sousa Jabbour, A. B., Jabbour, C. J. C., Godinho Filho, M., & Roubaud, D. (2018). Industry 4.0 and the circular economy: a proposed research agenda and original roadmap for sustainable operations. *Annals of Operations Research, 270*(1–2), 273–286. https://doi.org/10.1007/s10479-018-2772-8

Lüdeke-Freund, F., Gold, S., & Bocken, N. (2019). A review and typology of circular economy business model patterns. *Journal of Industrial Ecology, 23*(1), 36–61. https://doi.org/10.1111/jiec.12763

Mamalis, A. G., Spentzas, K. N., & Mamali, A. A. (2013). The impact of automotive industry and its supply chain to climate change: Somme techno-economic aspects. *European Transport Research Review, 5*(1), 1–10. https://doi.org/10.1007/s12544-013-0089-x

McDonough, W., & Braungart, M. (2003). Towards a sustainable architecture for the 21st century: the promise of cradle-to-cradle design. *UNEP Industry and Environment, 26*, 13–16.

Mhatre, P., Panchal, R., Singh, A., & Bibyan, S. (2021). A systematic literature review on the circular economy initiatives in the European Union. *Sustainable Production and Consumption, 26*, 187–202. https://doi.org/10.1016/j.spc.2020.09.008

Mont, O. (2002). Clarifying the concept of product–service system. *Journal of Cleaner Production, 10*(3), 237–245. https://doi.org/10.1016/S0959-6526(01)00039-7

Moreno, M., Turner, C., Tiwari, A., Hutabarat, W., Charnley, F., Widjaja, D., & Mondini, L. (2017). Re-distributed manufacturing to achieve a circular economy: A case study utilizing IDEF0 modeling. *Procedia CIRP, 63*, 686–691. https://doi.org/10.1016/j.procir.2017.03.322

Morfeldt, J., Davidsson Kurland, S., & Johansson, D. J. (2021). Carbon footprint impacts of banning cars with internal combustion engines. *Transportation Research Part D: Transport and Environment, 95*, 102807. https://doi.org/10.1016/j.trd.2021.102807

Mügge, J., Grosse Erdmann, J., Riedelsheimer, T., Manoury, M. M., Smolka, S.-O., Wichmann, S., & Lindow, K. (2023). Empowering end-of-life vehicle decision making with cross-company data exchange and data sovereignty via Catena-X. *Sustainability, 15*(9), 7187. https://doi.org/10.3390/su15097187

Mugge, R., Schoormans, J. P. L., & Schifferstein, H. N. J. (2005). Design strategies to postpone consumers' product replacement: The value of a strong person-product relationship. *The Design Journal, 8*(2), 38–48. https://doi.org/10.2752/146069205789331637

Murray, A., Skene, K., & Haynes, K. (2017). The circular economy: An interdisciplinary exploration of the concept and application in a global context. *Journal of Business Ethics, 140*(3), 369–380. https://doi.org/10.1007/s10551-015-2693-2

Niero, M., Jensen, C. L., Fratini, C. F., Dorland, J., Jørgensen, M. S., & Georg, S. (2021). Is life cycle assessment enough to address unintended side effects from Circular Economy initiatives? *Journal of Industrial Ecology, 25*(5), 1111–1120. https://doi.org/10.1111/jiec.13134

Parajuly, K., Fitzpatrick, C., Muldoon, O., & Kuehr, R. (2020). Behavioral change for the circular economy: A review with focus on electronic waste management in the EU. *Resources, Conservation & Recycling: X, 6*, 100035. https://doi.org/10.1016/j.rcrx.2020.100035

Parida, V., Sjödin, D., & Reim, W. (2019). Reviewing literature on digitalization, business model innovation, and sustainable industry: Past achievements and future promises. *Sustainability, 11*(2), 391. https://doi.org/10.3390/su11020391

Pearce, D. W., & Turner, R. K. (1990). *Economics of natural resources and the environment*. Johns Hopkins University Press.

Pieroni, M. P., McAloone, T. C., & Pigosso, D. C. (2019). Business model innovation for circular economy and sustainability: A review of approaches. *Journal of Cleaner Production, 215*, 198–216. https://doi.org/10.1016/j.jclepro.2019.01.036

Potting, J., Hekkert, M., Worrell, E., & Hanemaaijer, A. (2017). *Circular economy: Measuring innovation in the product chain*. PBL Netherlands Environmental Assessment Agency.

PWC. (2007). The automotive industry and climate change framework and dynamics of the CO2 (r)evolution. Retrieved June 19, 2023, from https://www.pwc.com/th/en/automotive/assets/co2.pdf

Reike, D., Vermeulen, W. J., & Witjes, S. (2018). The circular economy: New or Refurbished as CE 3.0?—Exploring controversies in the conceptualization of the circular economy through a focus on history and resource value retention options. *Resources, Conservation and Recycling, 135*, 246–264. https://doi.org/10.1016/j.resconrec.2017.08.027

Reim, W., Sjödin, D., & Parida, V. (2018). Mitigating adverse customer behaviour for product-service system provision: An agency theory perspective. *Industrial Marketing Management, 74*, 150–161. https://doi.org/10.1016/j.indmarman.2018.04.004

Revellio, F. (2022). Mythos: Langlebige Produkte sind schlecht fürs Geschäft. In A. Böckl, J. Quaing, I. Weissbrod, & J. Böhm (Eds.), *Mythen der Circular Economy* (pp. 31–45). https://doi.org/10.25368/2022.163

Ries, L., Beckmann, M., & Wehnert, P. (2023). Sustainable smart product-service systems: a causal logic framework for impact design. *Journal of Business Economics, 93*(4), 667–706. https://doi.org/10.1007/s11573-023-01154-8

Rockström, J., Steffen, W., Noone, K., Persson, A., Chapin, F. S., Lambin, E. F., Lenton, T. M., Scheffer, M., Folke, C., Schellnhuber, H. J., Nykvist, B., de Wit, C. A., Hughes, T., van der Leeuw, S., Rodhe, H., Sörlin, S., Snyder, P. K., Costanza, R., Svedin, U., … Foley, J. A. (2009). A safe operating space for humanity. *Nature, 461*(7263), 472–475. https://doi.org/10.1038/461472a

Rogers, H. A., Deutz, P., & Ramos, T. B. (2021). Repairing the circular economy: Public perception and participant profile of the repair economy in Hull, UK. *Resources, Conservation and Recycling, 168*, 105447. https://doi.org/10.1016/j.resconrec.2021.105447

Rusch, M., Schöggl, J.-P., & Baumgartner, R. J. (2022). Application of digital technologies for sustainable product management in a circular economy: A review. *Business Strategy and the Environment, 32*, 1159–1174. https://doi.org/10.1002/bse.3099

Santa-Maria, T., Vermeulen, W. J. V., & Baumgartner, R. J. (2022). How do incumbent firms innovate their business models for the circular economy? Identifying micro-foundations of dynamic capabilities. *Business Strategy and the Environment, 31*(4), 1308–1333. https://doi.org/10.1002/bse.2956

Scheepens, A. E., Vogtländer, J. G., & Brezet, J. C. (2016). Two life cycle assessment (LCA) based methods to analyse and design complex (regional) circular economy systems. Case: making water tourism more sustainable. *Journal of Cleaner Production, 114*, 257–268. https://doi.org/10.1016/j.jclepro.2015.05.075

Schöggl, J.-P., Stumpf, L., & Baumgartner, R. J. (2020). The narrative of sustainability and circular economy—A longitudinal review of two decades of research. *Resources, Conservation and Recycling, 163*, 105073. https://doi.org/10.1016/j.resconrec.2020.105073

Selvefors, A., Rexfelt, O., Renström, S., & Strömberg, H. (2019). Use to use—A user perspective on product circularity. *Journal of Cleaner Production, 223*, 1014–1028. https://doi.org/10.1016/j.jclepro.2019.03.117

Shevchenko, T., Saidani, M., Ranjbari, M., Kronenberg, J., Danko, Y., & Laitala, K. (2023). Consumer behavior in the circular economy: Developing a product-centric framework. *Journal of Cleaner Production, 384*, 135568. https://doi.org/10.1016/j.jclepro.2022.135568

Stahel, W. R. (2010). *The performance economy*. Palgrave Macmillan. https://doi.org/10.1057/9780230274907

Stahel, W. R. (2016). The circular economy. *Nature, 531*(7595), 435–438. https://doi.org/10.1038/531435a

Stahel, W. R. (2019). *The circular economy: A user's guide*. Routledge. https://doi.org/10.4324/9780429259203

Stumpf, L., & Baumgartner, R. J. (2022). Die Circular Economy—ein Konzept mit vielen Perspektiven. In A. Böckl, J. Quaing, I. Weissbrod, & J. Böhm (Eds.), *Mythen der Circular Economy* (pp. 5–12). https://doi.org/10.25368/2022.163

Suchek, N., Fernandes, C. I., Kraus, S., Filser, M., & Sjögrén, H. (2021). Innovation and the circular economy: A systematic literature review. *Business Strategy and the Environment, 30*(8), 3686–3702. https://doi.org/10.1002/bse.2834

Tecchio, P., McAlister, C., Mathieux, F., & Ardente, F. (2017). In search of standards to support circularity in product policies: A systematic approach. *Journal of Cleaner Production, 168*, 1533–1546. https://doi.org/10.1016/j.jclepro.2017.05.198

Tietze, F., & Hansen, E. (2017). To own or to use? Retrieved June 19, 2023, from https://doi.org/10.17863/CAM.13931

Tukker, A. (2004). Eight types of product–service system: eight ways to sustainability? Experiences from SusProNet. *Business Strategy and the Environment, 13*(4), 246–260. https://doi.org/10.1002/bse.414

Tukker, A. (2015). Product services for a resource-efficient and circular economy—a review. *Journal of Cleaner Production, 97*, 76–91. https://doi.org/10.1016/j.jclepro.2013.11.049

Valencia, A., Mugge, R., Schoormans, J., & Schiffersein, H. (2015). The design of smart product-service systems (PSSs): An exploration of design characteristics. *International Journal of Design, 9*(1).

Velenturf, A. P., & Purnell, P. (2021). Principles for a sustainable circular economy. *Sustainable Production and Consumption, 27*, 1437–1457. https://doi.org/10.1016/j.spc.2021.02.018

Wastling, T., Charnley, F., & Moreno, M. (2018). Design for circular behaviour: Considering users in a circular economy. *Sustainability, 10*(6), 1743. https://doi.org/10.3390/su10061743

Yang, M., & Evans, S. (2019). Product-service system business model archetypes and sustainability. *Journal of Cleaner Production, 220*, 1156–1166. https://doi.org/10.1016/j.jclepro.2019.02.067

Zhang, C., & Ji, W. (2019). Digital twin-driven carbon emission prediction and low-carbon control of intelligent manufacturing job-shop. *Procedia CIRP, 83*, 624–629. https://doi.org/10.1016/j.procir.2019.04.095

Zhang, W., & Xu, J. (2022). Advanced lightweight materials for automobiles: A review. *Materials & Design, 221*, 110994. https://doi.org/10.1016/j.matdes.2022.110994

Open Access This chapter is licensed under the terms of the Creative Commons Attribution 4.0 International License (http://creativecommons.org/licenses/by/4.0/), which permits use, sharing, adaptation, distribution and reproduction in any medium or format, as long as you give appropriate credit to the original author(s) and the source, provide a link to the Creative Commons license and indicate if changes were made.

The images or other third party material in this chapter are included in the chapter's Creative Commons license, unless indicated otherwise in a credit line to the material. If material is not included in the chapter's Creative Commons license and your intended use is not permitted by statutory regulation or exceeds the permitted use, you will need to obtain permission directly from the copyright holder.

6

Transforming Value Chains for Sustainability

Closing the Loop in the Age of Electromobility

Kai-Ingo Voigt, Lothar Czaja, and Oliver Zipse

6.1 Introduction

In the face of increasing global warming and extreme climatic conditions, 196 parties signed up to the Paris Agreement with the goal of limiting global warming to well below 2 °C compared with preindustrial levels, requiring net-zero emissions by 2050 (United Nations Framework Convention on Climate Change, 2022). The use of renewable energy and electromobility is essential for a transition to a carbon-free economy (Weimer et al., 2019). Current fossil-based road transport is the largest contributor to global warming within the transport sector, creating significant potential through the deployment of electric vehicles (Basia et al., 2021). Here, rechargeable lithium-ion batteries (also known as Li-ion batteries or LIBs) are currently the most favorable technological solution for the electric vehicle (EV) market (Weimer et al., 2019).

While EVs can offer several sustainability benefits, creating new and transforming existing automotive value chains to enable this transition is a formidable task. On the Road to Net Zero outlined in this book, *Transforming Value Chains for Sustainability*, thus marks a critical step that connects the previous chapter (Chap. 5) and the following chapter (Chap. 7). Chapter 5

K.-I. Voigt (✉) • L. Czaja
FAU Erlangen-Nürnberg, Nuremberg, Germany
e-mail: wiso-industry@fau.de

O. Zipse
BMW AG, Munich, Germany

introduced the general idea of the circular economy and its potential for *Creating Sustainable Products*. Chapter 6 now takes a deep dive into the EV battery value chains to review and discuss the complexity, potential, and challenges of what it means to strive to keep materials in a continuous cycle.

Since batteries and battery technologies are an essential part of modern electric vehicles, both the automotive value chain and the automotive battery industry must become a complex overall system in which the players' steps are interlocked and comprehensively regulated. At present, the Li-ion battery value chain still follows the approach of the traditional linear economy (Di Persio et al., 2020). In the context of meeting climate targets, the European Commission has also expressed the need for change in the battery industry. It commits to creating a competitive and sustainable battery value chain that adheres to circular economy principles, while developing high environmental and social standards. To achieve this, the battery production and recycling chains need to minimize their environmental footprint. Requirements for the safe and sustainable production, reuse, and recycling of batteries will play an essential role (Bielewski et al., 2021).

The purpose of this chapter is to provide an in-depth look at how the automotive industry's transition to electromobility is leading to far-reaching implications for the EV battery value chain. The chapter is divided into five sections. Section 6.2 sets the scene with a brief review of resource scarcity as a relevant strategic background for the circular economy. Section 6.3 then takes a detailed look at the different steps of the EV battery value chain, but without focusing on circularity yet. Section 6.4 presents the expert conversation between Prof. Oliver Zipse, Chairman of the Board of Management of Bayerische Motoren Werke (BMW) AG, and Prof. Dr. Kai-Ingo Voigt, Chair of Industrial Management at FAU Erlangen-Nürnberg. Section 6.5 returns to the EV battery value chain with a circularity perspective and discusses the technology and value chain steps for closing the loop in the EV battery life cycle. After giving an outlook on the challenges of circular EV battery value chains in Sect. 6.6, the chapter concludes in Sect. 6.7 with key takeaways and the link to the following chapter (Chap. 7) on *Sustainability in Manufacturing*.

6.2 In the Age of Resource Scarcity

The EV market is moving from a predominantly policy-driven market to one where organic customers are the most significant factor. In many countries, supply is a greater barrier to adoption than demand (BloombergNEF, 2022).

Based on the market size of electric mobility of 142 GWh in 2018, the battery market for EVs is expected to increase 16-fold in size by 2030, with a compound annual growth rate of 26.3% (World Economic Forum, 2019). These developments pose significant challenges to the industry, not only by covering material demand for vehicle production but also in proceeding with the vehicles after their end-of-life (EoL).

Regarding material demand, Germany (and thus the German industry, in particular) is almost entirely dependent on imports for fossil fuels, metallic raw materials, and many industrial minerals. There are many risk factors, ranging from political instability in some producing countries to strategic trade restrictions. In addition, companies are increasingly confronted with delivery difficulties, supply bottlenecks, and the risk of delivery disruptions. Increasing demand for raw materials from the developing and emerging countries is also leading to stronger competition on the raw materials market. This applies, in particular, to raw materials that are required for new technologies in the automotive industry, electronics, or environmental technology fields. High prices, price fluctuations, and supply bottlenecks are burdening the German economy. Companies are forced to diversify their sources of supply, hedge price risks, and substitute raw materials that are becoming scarcer (DIHK, 2022).

With regard to the battery market, which is particularly relevant for electromobility, the global battery market can be divided into primary and secondary batteries, with a ratio of 1 to 3. Whereas, in primary batteries, the chemical reaction is not reversible and the battery is designed only for a single use, the chemical reaction in secondary batteries is reversible. This reversible chemical process allows secondary batteries to be repeatedly charged and discharged. With a market share of almost 50% each, lead-acid and Li-ion batteries shared the global battery market for secondary batteries in 2019 (Zhao et al., 2021). The technical characteristics make Li-ion batteries particularly interesting for traction batteries in EVs. Although the basic principle is always the same, countless different Li-ion battery solutions are available, depending on the chemical composition and design.

The production of automotive Li-ion batteries uses many materials not previously required in the automotive sector. Moreover, battery use leads to six times higher mineral demand for electric vehicles than for conventional vehicles (International Energy Agency, 2018). This poses challenges for the industry regarding the continuous material supply of precious metals and rising demand (International Energy Agency, 2018). While some materials can be delivered without any problems, the so-called critical resources sometimes cause great difficulties.

Given the current trends and developments within battery chemistry, cobalt, graphite, lithium, manganese, and nickel are seen as critical battery raw materials and will be briefly presented (Bobba et al., 2020). Within the critical resources, cobalt, lithium, and graphite are assigned a further increased risk compared with nickel and manganese (Vereinigung der Bayerischen Wirtschaft, 2021).

Cobalt is mainly extracted as a by-product of copper and nickel mining. The Democratic Republic of Congo remains the leading source of mined cobalt as of 2021, accounting for 70% of global cobalt production. Subsequent processing occurs mainly in China, which has over 90% of the global refining and processing capacity (U.S. Geological Survey, 2022). This strong focus on mining and processing in two countries leads to a high risk for the supply chain of cobalt (International Energy Agency, 2021). China is the leading consumer of cobalt, with a strong focus of 80% on the rechargeable battery industry. There is an increasing trend to reduce the cobalt content within the battery chemistry (U.S. Geological Survey, 2022).

The security of the global lithium supply has recently become the highest priority of technology companies. Australia, Chile, and China account for 95% of the world production. The supply of two types of resources can be distinguished: the brine-based lithium sources from Chile and China and the spodumene ore from Australia. The type of resource also differentiates the subsequent processing and refining. China dominates the market in terms of hard-rock mineral refining facilities for spodumene ore, with 45% of total refining capacity. In contrast, 32% of the refining capacity is located in Chile and 20% in Argentina, with a focus on refining lithium from brine operations (U.S. Geological Survey, 2022). In the supply area, no major issue for the battery supply chain is found in the short- and medium-term future (Huisman et al., 2020). Despite recent developments in sodium-ion batteries, no large-scale material substitutes for lithium are expected in automotive batteries any time soon (U.S. Geological Survey, 2022).

Overall, 79% of the global graphite production is ensured by China, which accounts for one-quarter of the available amorphous graphite and three-quarters of the flake graphite (U.S. Geological Survey, 2022). China also dominates the downstream processing of spherical graphite. The graphite for Li-ion battery production has high requirements in terms of flake size and carbon content (Bobba et al., 2020). China therefore occupies a dominant position, and this strongly hinders any diversification of the supply chain. In addition to natural graphite, synthetic graphite powder and secondary synthetic graphite from machining graphite shapes have come increasingly to the fore (U.S. Geological Survey, 2022).

Indonesia, the United States, the Philippines, and Russia accounted for 75% of the world's nickel production in 2021 (U.S. Geological Survey, 2022). Li-ion batteries have high purity requirements for nickel and resort to nickel sulfate, which can be synthesized from Class 1 products with a purity of over 99.8% (International Energy Agency, 2021). Nickel already has a well-developed supply chain due to its versatile use in the past. Li-ion batteries comprise only a tiny part of the demand mix (International Energy Agency, 2018). Nevertheless, over the past 5 years, strong developments can be seen in the Asia/Pacific region (International Energy Agency, 2021). Here, Indonesia and the Philippines account for 50.7% of the global supply.

South Africa, Gabon, and Australia ensure the supply of manganese, providing 71.5% of the world production. No substitute is expected in Li-ion battery technology (U.S. Geological Survey, 2022).

In summary, in the age of resource scarcity, the supply of raw materials for electromobility, which will become increasingly important in the future, can be assumed to pose major challenges for the automotive value chain and trigger major change processes. In addition, the automotive industry is confronted with another major challenge: Even if electromobility is just picking up speed at present, researchers expect a huge annual volume of old battery returns by 2040. Concepts and techniques for the sustainable use of old Li-ion batteries are therefore just as much in demand as the value chains that are adapted and modified to meet these raw material challenges.

6.3   Value Chain Transformations

The automotive Li-ion batteries value chain spreads its process steps globally. The mining of materials, the following processing, and the batteries' production are distributed worldwide depending on availability, expertise, and production costs. While procuring critical raw materials is mainly located in the southern hemisphere, the subsequent processing and production of the cells occur in Asia. Usually, the final assembly of the modules and the EV battery takes place at the original equipment manufacturer (OEM), concentrated in Asia, the European Union (EU), and the United States. The single steps of the linear value chain can be divided into four phases: The extraction and procurement of materials with subsequent processing describe the upstream (Phase 1). In the midstream (Phase 2), the individual cell components are manufactured and assembled into a battery cell. The downstream (Phase 3) takes place at the OEM and includes the assembly of the battery cells into modules and packs, including their battery management system and auxiliary systems. This is followed by integrating the battery system into the electric vehicle. The end-of-life (Phase 4) describes the fourth phase, consisting of the removal from the EV and the subsequent second life or recycling of the valuable materials (Ciulla et al., 2021; Lebedeva et al., 2017). The first steps, including material procurement, processing, and component and cell production, are cost-driven; therefore, they are subject to global competition. Subsequently, the focus lies on the application and the specific customer requirements, which leads to a value orientation in the downstream area (Steen et al., 2017).

The high demand for Li-ion batteries is reflected noticeably in the upstream process step in the demand for raw materials. The raw materials required for Li-ion batteries are further subdivided into their criticality based on expected demand, natural occurrence, and production capacities. As discussed above, the literature defines cobalt, graphite, lithium, manganese, and nickel as critical materials. By nature, there is strong dependence on individual mineral-rich countries and regions, which leads to cooperation with countries with different labor conditions and standards of human dignity (Ciulla et al., 2021).

The extracted raw materials in their original form must be further processed and refined for use in production. Depending on the material, different purity and particle size requirements apply. These specific requirements must be met in order to be able to produce cell components. In 2020, the majority of the global processing capacity was in China (52%) and Japan (31%), highlighting the strong dependence on the Asian region (Bobba et al., 2020).

The subsequent midstream, starting with component production as the third step, is also dominated by China. Overall, 60% of manufacturing occurs in China, followed by Japan and Korea (Ciulla et al., 2021). Together, they cover around 85% of global component production, consisting of positive and negative electrodes, separators, electrolytes, and housing (Bobba et al., 2020). The valuable production of the electrodes consists of the successful coating of the carrier foil and follows a six-step process (Heimes et al., 2018).

Cell production describes the assembly of the components and is the fourth step of value creation. Like the previous one, this step is also strongly dominated by the Asian market. To minimize this regional domination, companies like BMW Group have already made decisions to locate and develop battery cell production in Europe and North America. The individual components are assembled into a battery cell representing the smallest unit. The assembly, including final finishing and testing, follows a seven-step process (Heimes et al., 2018). In general, the resulting production costs are divided into three phases: electrode production (39%), cell assembly (20%), and cell finishing (41%) (Kuepper et al., 2018).

The downstream is described by battery pack manufacturing and subsequent integration into the electric vehicle. This step takes place at the automotive OEM. For this purpose, several cells are combined to form modules, which are then bundled as a battery system. In addition to the modules, the battery system includes several mechanical and electrical components, such as housing, electronics, and a battery management system. The downstream phase concludes with the final assembly of the battery in the vehicle.

After reaching the minimum battery capacity and its end-of-life, the battery is removed from the vehicle. This is followed by the disposal or recycling of valuable components. Due to the increasing importance of this step for the fulfillment of a closed loop, the linear recycling process chain will be discussed in more detail in the following section.

6.3.1 Recycling of Lithium-Ion Batteries (LIBs)

The phases of the battery life cycle can mainly be divided into production, use, and recycling, including disposal (Fan et al., 2020). While the focus in the past was clearly on the first two phases, the latter will become increasingly important as the significance and demand for Li-ion batteries grow. The methods for dealing with LIBs are time-delayed due to the increase in battery demand; therefore, they must be established on an industrial scale. The increasing demand for raw materials can also be better met by additional recycling (Fan et al., 2020). Nevertheless, the recycling of LIBs is an emerging field that has not yet defined standardized and final processes (Neumann et al., 2022). This is also reflected in the literature, as most publications and research activities deal with necessary substeps within the recycling chain, but hardly examine the holistic chain with its supporting processes. The literature describes several approaches for future process steps concerning a holistic circular economy, but still shows considerable gaps between academic approaches and industrial reality (Neumann et al., 2022). The circular economy challenge has been identified as one of the pressing tasks and accelerating trends. The basis for the circular economy is the linear process flow for the recycling of spent Li-ion batteries on an industrial level.

In general, the process can be divided into four phases: The reverse logistics of the EV packs (Phase 1), the pretreatment of the EV packs to break them into enriched materials (Phase 2), the metallurgical treatment by recycling methods to preserve the specific materials (Phase 3), and the reintroduction of the pure materials into the market (Phase 4). The aim is to extract the valuable materials from the used batteries and return them to production. Current approaches focus mainly on recycling valuable and scarce materials mostly found in cathodes, such as cobalt, lithium, and nickel (Hua et al., 2020). In the future, the recycling of anodes and electrolytes should be included to increase the economic feasibility and sustainability of battery recycling. The necessary process steps are mostly academic approaches and far from industrial reality, but have gained increasing attention in recent years (Neumann et al., 2022). These developments are fundamental to ultimately speaking of recycling all parts and a holistic circular economy (Neumann et al., 2022).

6.3.2 Reverse Logistics (Phase 1)

The foundation for a successful and holistic recycling strategy is laid by reverse logistics (Voigt & Thiell, 2004), which is responsible for taking the used batteries out of circulation and transporting them to the subsequent recycling steps. The substeps of reverse logistics can be divided into material collection and sorting and transport and handling.

At present, no uniform and large-scale processes exist for collecting and sorting spent EV batteries. Standards and specifications are missing to enable the holistic recycling of all spent batteries in the future (Steward et al., 2019). In theory, the necessary steps are known and follow a simple sequence. The end-of-life vehicles must be collected as soon as the vehicles are taken out of service at the dealership or scrap yard. This is followed by transport to the disassembly plant, where they might be stored for some time. Here, the batteries are separated from the vehicle and collected (Steward et al., 2019). There are two main challenges at this stage: the heterogeneity in size and design and the difference in battery chemistries. To prevent a mix of materials and to increase the recycling efficiency of the subsequent metallurgical treatment, attention must be paid to ensure uniform battery chemistries. The lack of labels with essential information on the characteristics and composition of the batteries makes uniform sorting difficult, indicating that mandatory labeling will be essential in the future (Neumann et al., 2022).

The dismantled batteries are then transported to the recycling facilities for further processing. Due to the inherent dangers of Li-ion batteries, special safety requirements are imposed for further transport and handling. The hazards result from the high energy density and the toxic and flammable substances inside the battery. The greatest danger comes from thermal runaway, which is a cascade of uncontrolled exothermic reactions. This can be triggered by external heat sources, external and internal short circuits, or mechanical stresses and can lead to the ignition of the entire battery. For this reason, severe restrictions are placed on shipping quantities, safe packaging, size specifications, labeling requirements, and regulations for safety testing. These significantly affect transport costs, determined primarily by transport distance, transport volumes, capacity utilization, and additional safety precautions. On average, transport costs account for 41% of the total recycling costs and greatly influence the profitability of recycling. They also harm the balance sheet in terms of emissions, especially carbon dioxide (CO_2) emissions (Neumann et al., 2022).

6.3.3 Pretreatment (Phase 2)

The second phase comprises the pretreatment, intended to prepare the batteries for the material extraction process. Valuable components and materials must be separated and enriched based on differences in various physical properties (shape, density, and magnetic properties). Thus, higher recovery rates, lower energy consumption, fewer safety risks, and fewer environmental threats can be achieved. The pretreatment consists of a series of chemical and physical operations within the individual steps of discharging, disassembly, crushing, and separation (Hua et al., 2020).

The residual energy present in the spent batteries can lead to short circuits, resulting in explosions during the pretreatment process. The tiniest sparks can cause the ignition of volatile organic compounds during the crushing process that can lead to a fire. To counteract this danger to man and machine, batteries are first discharged and thereby stabilized (Neumann et al., 2022). Various industrial methods are available for discharging, with the brine method (salt-water-based baths) and the ohmic discharge method (controlled discharging via external circuits) being the most commonly used (Hua et al., 2020).

The second step in pretreatment is the disassembly of the EV packs. Here, the battery system is disassembled from the pack level to the module and cell levels. The aim is to achieve an initial rough presorting of the components to maximize economic benefits. First, the battery framework is opened, and the electrical connections between the components are cut. The mechanical connections between the components and the base are then released, and the electronic parts are removed. Finally, the spent battery cells are exposed. The lack of standards for the design and configuration of battery packs complicates any machine automation of the disassembly steps. Widely varying designs and configurations still require a high level of human involvement and manual handling (Hua et al., 2020).

Crushing is a further refinement of batteries at the cell level. In coarse shredding or fine crushing, the granularity of the materials is reduced for the subsequent processing steps. To reduce pollution and the risk of thermal runaway, the battery shredding or crushing processes can be carried out in an inert gas environment using carbon dioxide. Alternatively, crushing can be performed in a lithium brine to neutralize the electrolyte and avoid gas emissions (Hua et al., 2020).

The crushed materials, the so-called "black mass," are then separated in a multistage separation process. The main focus is the separation of the metallic particles (casing, copper, and aluminum foil) from the black mass. The latter consists of a mix of the active materials from the anode and cathode. It is the most valuable battery cell component and is to be maximally recovered in pretreatment (Neumann et al., 2022). The materials can be separated based on their differences in physical properties, such as size, density, ferromagnetism, and hydrophobicity. This is done in a multistage physical separation process consisting of multiple crushing and sieving steps, magnetic separation, and/or flotation (Hua et al., 2020).

6.3.4 Metallurgical Treatment (Phase 3)

The third phase of the recycling process describes the metallurgical treatment of the previously obtained enriched materials. For this purpose, the following metallurgical technologies are available, differing significantly in their design, properties, and degree of maturity: hydrometallurgy, pyrometallurgy, a mixture of both, biohydrometallurgy (bioleaching), and direct recycling (Hua et al., 2020). While the first two have already reached a sufficient level of technological maturity for industrial implementation, the latter two are still at the laboratory stage and have only demonstrated their technological feasibility under research conditions (Neumann et al., 2022).

Pyrometallurgical technology is based on the thermal treatment of spent batteries. A high-temperature furnace reduces the valuable metal oxides to a mixed alloy (Neumann et al., 2022). This process can be divided into three steps: preheating, plastic burning, and valuable metal reduction. The first two steps describe the thermal treatment, which first evaporates the electrolyte, thereby reducing the risk of explosion. This is followed by the burning of organic materials (e.g., plastics). Finally, at a temperature of 800–1000 °C, the materials are smelted and reduced to an alloy of valuable materials, such as copper, iron, cobalt, and nickel. The resulting slag contains lithium, aluminum, and calcium (Hua et al., 2020). Extensive pretreatment is not necessary.

Nevertheless, the output alloy must be posttreated and the materials preserved. Also, the slag should receive posttreatment to avoid discarding resources. The method is not economically interesting for batteries that do not contain cobalt and nickel (e.g., lithium ferrophosphate [LFP] batteries).

Hydrometallurgical technology is based on the leaching and extraction of valuable metals from spent LIBs using water-based solutions. The pretreated battery materials undergo a multistage process, with the following key procedures: leaching, precipitation, and solvent extraction (Hua et al., 2020). First, black mass is leached using mineral acids. The resulting leachate is precipitated of impurities to subsequently recover the valuable materials in a multistep solvent extraction process. By varying the pH of the acid used, manganese, cobalt, and nickel can be extracted successively in the form of salt mixtures. The final precipitation enables the lithium to be obtained as a salt mixture (Neumann et al., 2022).

The techniques of pyrometallurgy and hydrometallurgy can be combined to increase the recycling yield. The alloy resulting from pyrometallurgical treatments is refined using a hydrometallurgical process to isolate the metals. This allows a higher recovery rate for nickel and cobalt and increases the process robustness and flexibility to chemistry changes. However, this method does not solve the problem of slag, which remains unused as a waste product (Roland Berger, 2022).

Biohydrometallurgy uses microorganisms to recover valuable materials from spent batteries and offers a cost-efficient and eco-friendly alternative to the abovementioned approaches. As one of the biohydrometallurgical processes, bioleaching has gained a further attention in LIB recycling (Roy et al., 2021). Chemolithotrophic and acidophilic bacteria serve as the processing microorganisms. Iron ions and sulfur are energy sources used by these microorganisms to produce metabolites in the leaching medium (Moazzam et al., 2021). The microorganisms' activity produces organic and inorganic acids. These are applied to leach metals by converting the insoluble solids into soluble and extractable forms (Moazzam et al., 2021). They can dissolve several metals, such as cobalt, copper, lithium, manganese, and nickel. Nevertheless, this technology is still conducted only on a laboratory scale and is very time-consuming due to the time for cultivation of the microorganisms. After 10–15 days, the metals can be extracted with 80–95% efficiency (Roy et al., 2021).

Direct recycling recovers materials without affecting their original compound structure and decomposition (Hua et al., 2020). The fundamental idea lies in the refreshment and reactivation of active materials with still functional morphology. The capacity and properties lost through cycling can be restored, rather than breaking down active materials into their components for subsequent resynthesis. The methods used for this are still under research. They include thermal reactivation methods, hydrothermal relithiation, electrochemical methods for relithiation, short high-voltage pulses, and exposure to high lithium moieties, including re-sintering (Neumann et al., 2022). The active cathode material is recovered from the black mass without smelting or leaching (Roland Berger, 2022). Thus, the number of processing steps required to resynthesize the cathode materials can be reduced, lowering the environmental impact. It is currently the only process that enables economically viable LFP and lithium manganese oxide (LMO) cathode recycling. When selecting input materials, care must be taken to ensure uniform cathode chemistry. Like bioleaching, direct recycling is still limited to the laboratory scale, but it holds great potential for the future (Neumann et al., 2022).

6.3.5 Reintroduction into the Market (Phase 4)

After successfully extracting and recycling the pure materials, the raw materials can be put back into circulation in the fourth phase. They are again available as raw materials for new batteries and other products and must be distributed to the respective manufacturers.

With CO_2 emissions arising from both the manufacturing and recycling of batteries, the decarbonization of the automotive industry poses a cross-company challenge, as the vast majority of the ecological footprint is created in the supply chain. This creates the need to share emissions-related data across the value chain. Several digital solutions are currently emerging to address this need. With the ecosystem-based SiGreen approach for exchanging emissions data, Siemens developed a solution for efficiently querying, calculating, and passing on information about the real CO_2 footprint of products. This allows emission data to be exchanged along the supply chain and combined with the emission data from one's own value chain to create a real CO_2 footprint for products. This not only increases transparency in the automotive value chain, but also opens up new opportunities for making it more sustainable (Siemens, 2021). In the automotive industry, Catena-X emerges as a digital industrial data platform that allows OEMs and suppliers to share life-cycle-oriented data along the entire value chain (Catena-X, 2023). On the Road to Net Zero, Catena-X aims to establish standardized measurements to document real carbon data that reflect the real processes and location factors over the supply chain. In addition, Catena-X seeks to facilitate the data needed to improve traceability, efficiency, and circularity across value chain steps. As sustainability requires a transformation of entire industries, such digital ecosystems and new forms of data sharing will be crucial for fostering the value chains of the future.

6.4 Expert Conversation on Sustainability in the Supply Chain

Is Supply Chain Transparency the Key to Sustainability?

Voigt: Sustainability has gained strategic importance in the automotive industry, with significant relevance in its supply chains. BMW is a worldwide leader concerning sustainability aspects. Could you reflect on how you approach supply chain management (SCM) from a sustainability point of view? What are the major changes in the BMW supply chain network concerning digitalization, sustainability, or even the lessons learned from coronavirus disease (COVID-19)?

Zipse: The industry is right in the middle of society because our cars—our products—are very visible in the streets. Everyone knows the brand. There is a lot of discussion about sustainability. Society will not accept if you are not complying with specific regulations such as emissions or safety standards—regardless of whether you are in Europe, the United States, or Asia. Even worse: Your brand image and your value in the marketplace will immediately be diminished if you are not compliant, so you simply have to be. Otherwise, it would be very harmful. The diesel scandal is much more than a technical issue. It destroyed a lot of market value and future capital.

Voigt: How does BMW address this situation?

Zipse: BMW goes one step further. We put sustainability right into the core of our company policy and strategy. But that means we have to fulfill this promise. You must walk the talk. You cannot only say "I will do a statement" or "I will set a new target until 2050." It starts today! Otherwise nowadays the press will report that you say something but do something else.

Voigt: What role does the value chain play in this regard?

Zipse: Over 85% of our added value in the car is not manufactured or sometimes not even designed by us. It is designed by the supply chain and our partners. However, at the end of the day, you are responsible for aggregating all these supply chain components into a final product. So, there is a specific responsibility for any car manufacturer to be knowledgeable about the status of the supply chain, specifically when we talk about emissions, most prominently CO_2, and that you are aware of whom you give contracts. It is for this reason that we have implemented very close rules of conduct for all our supply chain members and rules about the transparency of what they are doing.

How to Establish Standards in a Contract Culture?

Zipse: Supply chains always have to deal with resource scarcity. That is of eminent importance for the future of any industry, and the automotive industry operates the largest supply chain in the world. What approaches can academia offer to improve managing resource scarcity in our supply chains?

Voigt: Every economic thought, every economic model, is centered around resource scarcity because we never have enough. We must decide what to use and what not to use. We see highly complex international value chain networks, especially in the automotive industry. As you mentioned, 85% of the value is created upstream, and you as the OEM have to measure this.

Zipse: How advanced are the measurement tools discussed in academia?

Voigt: Fortunately, our methods and systems are very effective in measuring economic impacts like costs and value created. You will likely know every cost in the whole value chain from every supplier. By contrast, we lack competence in measuring the social or ecological impact. That is the task of academia and practice to develop. It is not easy to do because economists have only one dimension—it is Dollar or Euro. Yet, in the ecological area, we have different emissions, so we need sensors for every possible emission that we have to measure. Subsequently, you have to react to these data wisely and therefore develop and implement decision systems. The head of procurement for Daimler cars, Dr. Güthenke, is working on a system using blockchain to measure the carbon footprint with every supplier. Do you have similar initiatives at BMW? How difficult is it to develop such a measuring system? How important is it to measure and acquire data to make decisions?

Zipse: BMW has more than 4000 direct suppliers and more than 12,000 suppliers overall. What kind of relationship do you have with them? Would you push responsibility to the supply chain and say: "You must be compliant with social standards and emission standards and cost contracts and so on." No. Delegating the problem is not the solution. The right approach is to do it together and support your suppliers. In many cases, they do not have the knowledge or economic power to implement specific steps, such as installing blockchain and reporting systems. The most important aspect is having a cooperative culture and not a contract culture. You need to nurture transparency over the supply chain.

Voigt: Could you give an example?

Zipse: Look at the scarcity of automotive chips. If we had more transparency and reliability about the n-tier supply chain, such as knowing where our orders are or the capacity status in factories, the problem would be diminished and much easier to solve because it is based on data.

Voigt: Interesting example! I see the huge potential of data sharing across the value chain.

Zipse: Yes. But supply chain partners will only be enticed to share data with you if they trust you. Of course, we have tremendous technical possibilities, like blockchain. But look at what could happen with GAIA-X: Instead of the Internet of things, you could have the Internet of companies. The technical possibility is there. The rest is trust. This is a great opportunity.

How Can Suppliers Contribute?

Voigt: If you come up to any of your 4000 direct suppliers telling them "Your EMAS certification and ISO 14001 is not enough. You must do more in this area.", they will react "OK, but then we have a lot more costs and no benefit." But we know that the suppliers' contribution is needed. Otherwise, we cannot reach our goals. The situation was quite similar back when we introduced quality systems: Suppliers were unhappy, but eventually, they did it. What is your message to key suppliers?

Zipse: Doing everything yourself is not impactful enough. If you have contracts with suppliers, it is good if your fellow competitors do the same. When we have sustainability clauses in our contracts, we try to ensure that they are similar within the automotive industry. That is very important because then the supplier has no choice. Look at electric cars. The biggest CO_2 footprint comes from the supply chain, 60% just from the sourcing and production of the battery cell, as opposed to the internal combustion engine, where it comes from the life cycle of the car.

Voigt: So how do you reduce the overall footprint of electric cars?

Zipse: All of our five battery cell suppliers have to comply and be transparent about where their energy source comes from—it is all 100% green energy. It is a matter of contract negotiations. When you put that into a contract, you assume it will cost more. But this is your proof that you have a functioning sustainability strategy in place. Suppliers understand that this is very important. So, in the long run, this will be the industry standard.

How Can Autonomous Driving Improve Sustainability?

Voigt: Academia and research prove that in the long run everybody has an advantage. Let's move to another topic: Do you think autonomous driving can bring ecological advantages?

Zipse: The basic question is "How is traffic organized?" Solving traffic problems is not the goal of our industry. But improving the situation so that you have fewer traffic jams is our concern. If you look at certain cities like Los Angeles or Chinese cities, the average driving speed is 20 km/h. The car is standing more than it is driving.

Voigt: What can technology offer to address this challenge?

Zipse: Self-driving cars offer an improvement when automated driving allows them to find their own way through very complex traffic situations. There will be an effect if we organize this in a good way. In addition to the impact on traffic, the productivity and well-being level of passengers will be increased. Even sleeping could be a productivity measure. Autonomous driving, even Level 2+, where you can take your mind off driving a little bit, would be a big step forward.

Do you have any scientific evidence of how autonomous driving could improve sustainability?

Voigt: I have been working in the automotive industry academically for 20 years, but I have not conducted any research on this topic myself, nor have I read any market study on how car-buyers are reacting to this. In all the discussions about electromobility and autonomous driving, the customer is barely mentioned. That is surprising, because a company has to produce products and services for customers, and the customer's desire should be the starting point.

How Can Customer Needs Be Integrated in the Process?

Zipse: The strategy of any product-driven company starts with the customer. The goal is to connect the customer with other stakeholder interests. Today's customer is very much aligned with societal goals. A car that obviously does not contribute to social and environmental aspects will have a reduced market share—even though there will always be some kind of niche.

Voigt: So you see the customer as potential drivers for sustainability?

Zipse: If you have an attractive product that also contributes to sustainability, it will have great market access. Look at sneakers: A fully recyclable sneaker is more expensive than a regular sneaker. You can create extremely attractive products and at the same time integrate sustainability and impact into the process. It is not a contradiction. Quite the opposite: If you neglect the customer, you are not even contributing.

Voigt: From the concept of market diffusion, we know that we do not have the whole market from the very beginning. Opening the market and developing the market are tasks of successful innovation management. How do you develop the market?

Zipse: BMW is proud to be a pioneer for driving dynamic cars, but we also integrate sustainability into our strategy. Customers are surprised that this integration works. You do not give something up to gain something else; you integrate it into a strategy. Only financially very successful companies have the resources to take the next step. Capital markets these days are very much linked to sustainability targets. It is again no contradiction. Capital markets demand more than just financial targets. Take the Taxonomy or the Corporate Sustainability Reporting regulations in the EU. Such frameworks are becoming more and more a part of your reporting.

Can Industry 4.0 Contribute to a More Sustainable Future?

Voigt: Let us move on to Industry 4.0. I was pleased to lead a research project at the University of Erlangen-Nuremberg to investigate whether Industry 4.0 can be a concept for more sustainable value creation in the industry. The automotive industry has always been an innovator in production technologies, so you could even do without Industry 4.0. But what is BMW's strategy with regard to Industry 4.0? Do you see any sustainability benefits from using Industry 4.0?

Zipse: Industry 4.0 is a German brand. If you go to Japan or the United States, they will say, "Industry 4.0? That comes from Germany." It is the combination of digitalization of factories with production systems, efficiency, waste reduction, and so on. Industry 4.0 and the Internet of things have made a huge step forward in terms of profitability. Maybe that would be an interesting area of research. How big was the actual cost reduction per unit through Industry 4.0? You see it in all factories.

Voigt: So do you see differences between Germany or Europe and the rest of the world?

Zipse: The most interesting thing is to look at large-scale digitalization in different application contexts. On the retail side, you would immediately assume that the American companies, the big-scalers, are leading the pack. On the industrial side, it is exactly the opposite. Europe, and Germany in particular, is leading the world in terms of value creation through Industry 4.0 in the industrial context—we see this through our own experience and research. In our factories, Industry 4.0 is the standard, which has a big impact on sustainability issues. Now, we are going to the next step: The Internet of companies—let's call it Industry 5.0.

Voigt: I see the link to our value chain discussion…

Zipse: Absolutely. We have already touched upon it: A supply chain is an Internet of companies. To create transparency in the supply chain, we need an Internet of companies. We have an automotive alliance that we founded together with SAP, Bosch, ZF, and also Continental. In this alliance, Catena-X, we have decided to build platforms to connect companies with each other, under specific safety rules. We want to make that a contribution to the broader EU policy concept of GAIA-X.

Voigt: We surveyed about 500 major industrial companies, and cost reduction was a key benefit of Industry 4.0. Yet, Industry 4.0 is not only a way to reduce costs, but in the electronics industry, for example, all the standard products can be transformed into more customer-oriented products. You could deliver millions of variants of each car. Do you see such benefits of Industry 4.0 beyond cost reduction?

Zipse: Of course. These benefits include not only cost savings but also quality improvements and your enhanced ability to manage complexity. But every transformation demands an initial investment. And we are getting there. Especially the Internet of tools is becoming a standard more and more. If you buy a new factory machine, it is already part of this world. Germany, but also Europe as a whole, is developing amazing applications based on the Internet of things, especially when practice and academia work together. I appreciate that you are doing this research in the application world. Another important research topic could be to define the next step in supply chain management in terms of creating transparency for regulatory purposes, quality purposes, cost purposes, and, of course, sustainability.

Voigt: We are willing to do this research, but we need practical partners to create value.

6.5 Closing the Loop

Coming back to the manufacturing and recycling of batteries, the linear value chain for Li-ion batteries is currently the dominant approach in the industry (Di Persio et al., 2020). Recent developments regarding future demand and supply, sustainability, and compliance with climate targets require closing the linear chain to a closed loop. Thus, the circular economy approach will have to be pursued, which is inevitable for future sustainable development. The circular economy is an economic system based on avoiding waste and promoting the continuous use of resources rather than sourcing new materials in the current linear economy. It focuses on waste management and aspects related to material reduction, reuse, recycling, and responsible manufacturing. It aims to develop new industries and jobs, reduce emissions, and increase efficiency in the use of natural resources.

In the transportation and power sectors, the circular economy is seen as a significant near-term driver of compliance with the Paris Agreement on climate change. The closed-loop approach would allow for a 30% reduction in CO_2 emissions from these sectors (Zhao et al., 2021). In the near future, a large number of Li-ion batteries will be retired and become part of the waste stream (Hua et al., 2020). To maximize the value of end-of-life batteries, they will be reused in various forms, such as remanufacturing and repurposing into new systems. In the final step, the valuable materials are to be extracted through recycling in order to be returned to the initial steps of the cycle (Hua et al., 2020).

The stages of the battery life cycle in a circular economy, and thus the sequence of steps in the value chain, consist of two interrelated cycles. First, the primary life cycle includes all steps up to the use of the battery in the vehicle and ends with recycling. In addition, the secondary life cycle will become increasingly important, which describes the reuse of the used EV batteries in new applications, the so-called "second life" (Gernant et al., 2022). This combination is intended to achieve the maximum yield from the materials and efforts expended, thereby reducing the relative resource consumption and emissions over the life cycle and maximizing the return on carbon investment incurred to produce it (Niese et al., 2020). Regardless of whether a battery has only completed the first life cycle or also through the second life cycle, the recycling of the batteries and thus the extraction of valuable materials close the circle.

The primary life cycle is initially characterized by the substeps already known from the linear value chain. Strictly speaking, the closed loop does not allow the process steps to be divided into upstream, midstream, and downstream anymore. However, the respective substeps are still reflected in the circular economy. The upstream consists of the extraction and processing of raw materials. This is followed in the midstream by the production of the individual cell components and their subsequent completion as finished cells. Finally, in the downstream, the battery pack is manufactured by the OEM and then installed in the EV. The completion of vehicle production marks the beginning of the first utilization phase of the battery in the EV. The total range of an EV is reported to be between 120,000 km and 240,000 km, with most manufacturers guaranteeing a range of around 160,000 km and a lifetime of 8 years (Hua et al., 2020). As usage increases and capacity losses occur, LIBs can no longer meet performance and energy requirements, such as driving range and acceleration (Hua et al., 2020). This is reflected in the battery's state of health, which typically reaches end-of-life at a capacity loss of 20–30%. Even during initial use, degraded or defective battery modules can be replaced with end-of-life modules as part of reconditioning and repair to further utilize the capacity of the remaining modules. Due to homogeneous battery aging resulting from more mature technologies and battery management systems, reconditioning will be limited to only 5% of end-of-life batteries in the long term (Zhao et al., 2021). Based on the analysis and the characteristics of the battery, it must be decided whether the battery will be part of the secondary life cycle and thus of the second use or whether it will be directly part of the recycling step.

The secondary life cycle and its applications focus on the value of repurposing a partially used battery, as opposed to subsequent recycling, which focuses on the value of the battery's metal content (Niese et al., 2020). The sequence of steps follows battery screening, battery disassembly and reassembly, and the subsequent application of repurposed batteries (Shahjalal et al., 2022). The technical feasibility of the battery chemistry and the associated economic viability of the second life are fundamental to the secondary life cycle. This consideration takes place after the first life cycle in reverse logistics and analytics as part of a precise suitability test. Methods such as electrochemical impedance spectroscopy, current interruption analysis, and capacity analysis are used (Kehl et al., 2021). The predominant use of used Li-ion batteries is in energy storage systems (ESSs). In addition, they can be used to refurbish and repair defective first-life battery modules. Repurposed Li-ion batteries will become increasingly important in sectors such as microgrids, smart grids, renewable energy, and area and frequency regulation. Specifically, they can be used in stationary grid applications, off-grid stationary applications, and mobile applications (Shahjalal et al., 2022). In particular, the increasing integration of renewable energies into the energy grid will boost the demand for stationary energy storage systems. They allow balancing between the irregularity of renewable energy generation with demand deviations and act as a buffer for grid stabilization (Shahjalal et al., 2022). The requirements for batteries in EVs differ from those in ESS, especially regarding cycling stability, power density, cooling, shock resistance, and safety. The requirements for ESS are significantly lower and easier to meet than those for EVs. Factors such as power density and shock resistance are less relevant than before. Differences can also be seen in the individual battery chemistries. Low-cost cell chemistries, in particular, seem to be more attractive for the second life, as they are technically more feasible and less interesting for direct recycling due to less expensive cell materials. LFPs, for example, have higher cycle stability, intrinsic safety, and lifetime than high-end technologies. The end-of-life in the second use occurs when a health state of 40–50% is reached. Subsequently, the materials should be extracted in the final recycling step and added to the beginning of the cycle (Gernant et al., 2022).

6.6 Outlook and Further Challenges

The challenges of the future automotive battery value chain are seen in the overarching issues of the battery industry as well as in further subcategories. With the introduction of autonomous driving, the classic value creation system in the automotive industry is seen in danger and significant disruptions are expected, especially in customer–OEM business relationships and ownership models. Nonautomotive players, such as Google, Waymo, Huawei, and Apple, are seen as disruption drivers. In general, (technical) challenges are expected in all areas of the automotive battery value chain. These are complemented by the importance of economies of scale, whose influence will increase sharply in the future. To be economically attractive, any future technology will require a high degree of standardization on the material side and in the cell format (shape and design). In particular, the need for standardization will increase as soon as it is considered from a total cost of ownership model perspective. In addition, the cost of battery technology in general will remain a challenge. This is primarily due to manufacturing, production processes, and raw materials. The need to balance user requirements with the cost of battery technology will be another challenge. To reach the mass market and mainstream electrification, many technology points still need to be improved to reduce costs. Apart from the battery, the development of the electrical infrastructure, including charging speed, is also seen as a key challenge for successful implementation.

Several experts see the circular economy of battery technology as a key challenge. This starts with the visibility of the batteries. Within the EU, the car manufacturers are legally responsible for the battery once it has reached its end-of-life (EoL) stage. To ensure this, they should always know where their EoL battery is located. This overview is significantly complicated in today's widespread classic car ownership model and is still an unsolved problem. The development of a comprehensive data infrastructure with information about the vehicle's current position in the value chain is becoming inevitable, in view of the increasing number of vehicles. To date, the foundations for this are lacking; the first step in this direction is the introduction of standardized battery passports and a digitally networked value chain that includes all relevant suppliers and partners. Furthermore, a closed material cycle for batteries and the necessary materials is perhaps the most crucial point for establishing the value chain in the long term. Procuring the necessary materials for market ramp-up should not cause any problems currently. However, in the long term, beyond 2050, the system is unlikely to work without an almost 100% closed-loop economy. For this, the cycle must be closed, and interfaces must be established. The question of who will be responsible for the division, one player for the entire cycle or different players, still needs to be clarified and increases the relevance of the intersections. Some experts address the degree of circularity and emphasize its importance in meeting carbon intensity and environmental impact expectations. Many projections for reducing the carbon footprint of battery production are based on the use of recycled materials. To meet the expected levels, experts see strong political action as imperative.

The material chain describes another challenge. The supply of resources and raw materials is a weak point and represents a major challenge in Europe, which requires a more sovereign positioning concerning its dependence. As a solution, a more sustainable design of the established supply chains and efforts to enter into partnerships with other countries are discussed. Even if the dependency cannot be resolved, Europe should try to adapt the value chain conditions to its sustainability vision and ideals. It should promote a sustainable value chain design around the extraction and processing of resources and pay attention to working and social conditions. Furthermore, changes in battery technologies are expected to have a significant impact on the material chain. These will lead to a change in material requirements, for example, with the decreasing demand for cobalt, the increasing demand for manganese, and the trend toward LFP chemistry. The shift to solid-state technology and metallic anodes will also overturn the current situation.

Expert opinions diverge in the area of capacity-building. While some experts believe that there is no problem in scaling up and meeting battery demand as long as sufficient raw materials are available, others see substantial challenges in building up production capacity and the associated need for materials. They also mention the current strategic planning conflict on capacity building. Decisions to build battery manufacturing and recycling capacity, in terms of location and battery chemistries, and to cooperate with energy storage system operators, must be made now so that sufficient capacity will be available a decade from now. This leads to the problem that many strategic decisions must be made based under uncertainty.

The production processes represent a further challenge. The robustness of all raw material and material processing synthesis processes is considered to be sufficiently high, as experience from the fast-moving consumer goods sector can be passed on here. The situation is different for innovations in the process steps, where uncertainties arise for the next-generation batteries regarding how raw materials or precursors for syntheses can be produced on a large scale. The same applies to the production of cell components and cells, for which there are no empirical values from large-scale industrial handling, highlighting the lack of technology and the need for technology development. On the production side, the processing of the solid-state electrolyte and the metallic anodes are seen as major issues. While some subprocesses of the next-generation batteries, such as dry pressing, are already at a medium level of maturity on an industrial scale, many other steps, especially in assembly, still pose significant challenges. Moreover, experts see significant cost reduction potential in establishing a dry coating process, which is still a complex process with high labor and energy costs. In the future, they see water-based processes with no solvents.

And as if that was not enough, experts see reverse logistics as another challenge. The difficulty of returning the EoL batteries is evident in the entire organization of the logistics chain for second life and recycling. It requires holistic cooperation between established and new players who have not yet worked together to this extent. The complexity is also reflected in the logistics costs. In the EU, batteries are classified as hazardous goods, requiring many obligations, certificates, and agreements for their transport. Due to the different implementation of regulations in the EU countries, country-specific adaptation and verification of the transport are required. This makes the transportation of batteries a slow and an expensive process. Even further challenges are seen in the collection of EoL batteries. Although the visibility of automotive batteries at the end-of-life is higher than for batteries from consumer goods, the different possibilities for second-life applications make highly efficient and high-quality recycling of critical materials still challenging. A clear separation of battery chemistries is necessary to ensure high quality and clean recycling. The problem of classification of EoL batteries is still unresolved. This requires information from the OEMs, which is currently difficult to obtain.

The final challenge lies in recycling and the revision of current recycling processes. Many difficulties and unresolved issues are currently seen here, with a clear gap between recyclers and producers. The current recycling processes are seen as inefficient and misrepresented. Recyclers often simply shred batteries and dispose of the so-called black mass in landfills, with no recovery and processing of raw materials. However, even companies that do recover raw materials use processes that call for further improvements. The established recycling processes are not the most efficient because they require a great deal of energy and cost to break everything down. What is needed instead is the development of gentler recycling methods. The problem currently lies in the small scales and the heterogeneity of cell chemistries. A process can only be properly optimized when the defined cell chemistries with expected materials are available. Another threat lies in the increasing popularity of LFP batteries. Due to the excellent availability of materials and their low cost, these batteries are becoming increasingly popular for nonpremium vehicles. At the end of their service life, in 10 to 15 years, many LFP batteries will be available that no one wants to recycle due to their lack of valuable materials and economic calculations. Companies and governments are not attacking the issue of LFP recycling. It is up to the government to implement regulatory policies to incentivize the recycling of LFPs.

6.7 Conclusion

The transition to a carbon-free economy is essential to limit global warming, and using renewable energy and electromobility is critical for achieving this. In the automotive industry, however, the transition to EVs shifts carbon footprint considerations upstream. Understanding, managing, and innovating the value chains of the future are therefore key on the Road to Net Zero.

How can sustainable value chains for the future be developed?—We would like to highlight five takeaways from this chapter that invite further discussion:

1. Reducing negative ecological and social impacts, not only for electric vehicles but more generally, requires a value chain perspective and circular thinking.
2. While closing the loop of material flows offers huge potential for meeting carbon reduction and environmental impact expectations, there are significant technological, organizational, and regulatory barriers at each step of the circular value chain.
3. Since circularity requires adequate data, further digitization such as the digital battery passport and adequate forms of data sharing are needed.
4. No single company can address the challenges of circular value chains alone. Instead, collaboration is needed, both along value chains and within industries.

5. The consistent further development of battery storage technologies can make a further decisive contribution to counteracting the prevailing scarcity of resources and will significantly influence the design of future sustainable supply chains.

On the Road to Net Zero, value chains thus play a pivotal role. Despite the importance of the upstream value chain, however, the core activities of industrial OEMs still lie in their own manufacturing processes. This is where the various components of complex supply networks are assembled into valuable products. Moreover, manufacturing is where companies have the greatest degree of control and can directly address their environmental footprint. For this reason, the following chapter (Chap. 7) now looks at *Sustainability in Manufacturing*.

References

Basia, A., Simeu-Abazi, Z., Gascard, E., & Zwolinski, P. (2021). Review on state of health estimation methodologies for lithium-ion batteries in the context of circular economy. *CIRP Journal of Manufacturing Science and Technology, 32*, 517–528. https://doi.org/10.1016/j.cirpj.2021.02.004

Roland Berger (Ed.). (2022). The Lithium-Ion (EV) battery market and supply chain: Market drivers and emerging supply chain risks. Retrieved June 26, 2023, from https://content.rolandberger.com/hubfs/07_presse/Roland%20Berger_The%20Lithium-Ion%20Battery%20Market%20and%20Supply%20Chain_2022_final.pdf

Bielewski, M., Blagoeva, D., Cordella, M., Di Persio, F., Gaudillat, P., Hildebrand, S., Mancini, L., Mathieux, F., Moretto, P., Paffumi, E., Paraskevas, D., Ruiz, V., Sanfélix, J., Villanueva, A., & Zampori, L. (2021). *Analysis of sustainability criteria for lithium-ion batteries including related standards and regulations*. Publications Office of the European Union. https://doi.org/10.2760/811476

BloombergNEF (Ed.). (2022). *Electric vehicle Outlook 2022: Executive summary*. Retrieved June 26, 2023, from https://about.newenergyfinance.com/electric-vehicle-outlook

Bobba, S., Carrara, S., Huisman, J., Mathieux, F., & Pavel, C. (2020). *Critical raw materials for strategic technologies and sectors in the EU: A foresight study*. Publications Office of the European Union. https://doi.org/10.2873/58081

Catena-X. (2023). *The Vision of Catena-X*. Retrieved July 9, 2023, from https://catena-x.net/en/vision-goals

Ciulla, F., Dayal, N., & Gujral, A. (2021). Charged-up demand brings challenges to the battery value chain. L.E.K. Consulting, 2021(23). Retrieved June 26, 2023, from https://www.lek.com/insights/ei/charged-demand-brings-challenges-battery-value-chain

Deutscher Industrie- und Handelskammertag (DIHK) (Ed.). (2022). Versorgung mit Rohstoffen. Retrieved June 26, 2023, from https://www.dihk.de/de/themen-und-positionen/wirtschaftspolitik/rohstoffe/versorgung-mit-rohstoffen-6254

Di Persio, F., Huisman, J., Bobba, S., Alves Dias, P., Blengini, G. A., & Blagoeva, D. (2020). *Information gap analysis for decision maker to move EU towards a circular economy for the lithium-ion battery value chain.* Publications Office of the European Union. https://doi.org/10.2760/069052

Fan, E., Li, L., Wang, Z., Lin, J., Huang, Y., Yao, Y., Chen, R., & Wu, F. (2020). Sustainable recycling technology for Li-ion batteries and beyond: Challenges and future prospects. *Chemical Reviews, 120*(14), 7020–7063. https://doi.org/10.1021/acs.chemrev.9b00535

Gernant, E., Seuster, F., Duffner, F., & Jambor, M. (2022). Understanding the automotive battery life cycle: A comprehensive analysis of current challenges in the circular economy of automotive batteries. Porsche Consulting. Retrieved June 26, 2023, from https://newsroom.porsche.com/dam/jcr:5a063b1d-7d12-4072-94ee-e4c479cd1621/Understanding%20the%20Automotive%20Battery%20Life%20Cycle_C_Porsche%20Consulting_2022.pdf

Heimes, H. H., Kampker, A., Lienemann, C., Locke, M., Offermanns, C., Michaelis, S., & Rahimzei, E. (2018). *Lithium-ion battery cell production process* (3rd ed.). PEM der RWTH Aachen University; DVMA.

Hua, Y., Zhou, S., Huang, Y., Liu, X., Ling, H., Zhou, X., Zhang, C., & Yang, S. (2020). Sustainable value chain of retired lithium-ion batteries for electric vehicles. *Journal of Power Sources, 478*, 228753. https://doi.org/10.1016/j.jpowsour.2020.228753

Huisman, J., Ciuta, T., Mathieux, F., Bobba, S., Georgitzikis, K., & Pennington, D. (2020). *RMIS–raw materials in the battery value chain.* Publications Office of the European Union. https://doi.org/10.2760/239710

International Energy Agency (Ed.). (2018). Global EV Outlook 2018: Towards cross-modal electrification. Retrieved June 26, 2023, from https://iea.blob.core.windows.net/assets/387e4191-acab-4665-9742-073499e3fa9d/Global_EV_Outlook_2018.pdf

International Energy Agency (Ed.). (2021). The role of critical minerals in clean energy transitions: World energy outlook special report. Retrieved June 26, 2023, from https://iea.blob.core.windows.net/assets/ffd2a83b-8c30-4e9d-980a-52b6d9a86fdc/TheRoleofCriticalMineralsinCleanEnergyTransitions.pdf

Kehl, D., Jennert, T., Lienesch, F., & Kurrat, M. (2021). Electrical characterization of Li-ion battery modules for second-life applications. *Batteries, 7*(2), 32. https://doi.org/10.3390/batteries7020032

Kuepper, D., Kuhlmann, K., Wolf, S., Pieper, C., Xu, G., & Ahmad, J. (2018). *The future of battery production in electric vehicles.* Boston Consulting Group. Retrieved June 26, 2023, from https://www.bcg.com/publications/2018/future-battery-production-electric-vehicles

Lebedeva, N., Di Persio, F., & Boon-Brett, L. (2017). *Lithium ion battery value chain and related opportunities for Europe*. Publications Office of the European Union. https://doi.org/10.2760/6060

Moazzam, P., Boroumand, Y., Rabiei, P., Baghbaderani, S. S., Mokarian, P., Mohagheghian, F., Mohammed, L. J., & Razmjou, A. (2021). Lithium bioleaching: An emerging approach for the recovery of Li from spent lithium ion batteries. *Chemosphere, 277*, 130196. https://doi.org/10.1016/j.chemosphere.2021.130196

Neumann, J., Petranikova, M., Meeus, M., Gamarra, J. D., Younesi, R., Winter, M., & Nowak, S. (2022). Recycling of lithium-ion batteries - current state of the art, circular economy, and next generation recycling. *Advanced Energy Materials, 17*, 2102917. https://doi.org/10.1002/aenm.202102917

Niese, N., Pieper, C., Arora, A., & Xie, A. (2020). *The case for a circular economy in electric vehicle batteries*. Boston Consulting Group. Retrieved June 26, 2023, from https://www.bcg.com/de-de/publications/2020/case-for-circular-economy-in-electric-vehicle-batteries

Roy, J. J., Srinivasan, M., & Cao, B. (2021). Bioleaching as an eco-friendly approach for metal recovery from spent NMC-based lithium-ion batteries at a high pulp density. *ACS Sustainable Chemistry & Engineering, 9*(8), 3060–3069. https://doi.org/10.1021/acssuschemeng.0c06573.s001

Shahjalal, M., Roy, P. K., Shams, T., Fly, A., Chowdhury, J. I., Ahmed, M. R., & Liu, K. (2022). A review on second-life of Li-ion batteries: Prospects, challenges, and issues. *Energy, 241*, 122881. https://doi.org/10.1016/j.energy.2021.122881

Siemens. (2021). *Siemens entwickelt ökosystembasierten Ansatz für den Austausch von Emissionsdaten*. Retrieved June 26, 2023, from https://press.siemens.com/global/de/pressemitteilung/siemens-entwickelt-oekosystembasierten-ansatz-fuer-den-austausch-von

Steen, M., Lebedeva, N., Di Persio, F., & Boon-Brett, L. (2017). *EU competitiveness in advanced Li-ion batteries for e-mobility and stationary storage applications - opportunities and actions*. Publications Office of the European Union https://doi.org/10.2760/75757

Steward, D., Mayyas, A., & Mann, M. (2019). Economics and challenges of li-ion battery recycling from end-of-life vehicles. *Procedia Manufacturing, 33*, 272–279. https://doi.org/10.1016/j.promfg.2019.04.033

U.S. Geological Survey (Ed.). (2022). *Mineral commodity summaries 2022*. U.S. Geological Survey. https://doi.org/10.3133/mcs2022

United Nations Framework Convention on Climate Change (Ed.). (2022). *The Paris Agreement: What is the Paris Agreement?* Retrieved June 26, 2023, from https://unfccc.int/process-and-meetings/the-paris-agreement/the-paris-agreement

Vereinigung der Bayerischen Wirtschaft (Ed.). (2021). Rohstoffsituation der bayerischen Wirtschaft. Retrieved June 26, 2023, from https://www.vbw-bayern.de/Redaktion/Frei-zugaengliche-Medien/Abteilungen-GS/Wirtschaftspolitik/2020/Downloads/201202-Studie-Rohstoffe.pdf

Voigt, K.-I., & Thiell, M. (2004). Industrielle Rücknahme- und Entsorgungssysteme. In G. Prockl, A. Bauer, A. Pflaum, & U. Müller-Steinfahrt (Eds.), *Entwicklungspfade und Meilensteine moderner Logistik* (pp. 389–418). Gabler Verlag. https://doi.org/10.1007/978-3-322-89044-3_19

Weimer, L., Braun, T., & vom Hemdt, A. (2019). Design of a systematic value chain for lithium-ion batteries from the raw material perspective. *Resources Policy, 64*, 101473. https://doi.org/10.1016/j.resourpol.2019.101473

World Economic Forum (Ed.). (2019). A vision for a sustainable battery value chain in 2030: unlocking the full potential to power sustainable development and climate change mitigation. Retrieved June 26, 2023, from https://www3.weforum.org/docs/WEF_A_Vision_for_a_Sustainable_Battery_Value_Chain_in_2030_Report.pdf

Zhao, Y., Pohl, O., Bhatt, A. I., Collis, G. E., Mahon, P. J., Rüther, T., & Hollenkamp, A. F. (2021). A review on battery market trends, second-life reuse, and recycling. *Sustainable Chemistry, 2*(1), 167–205. https://doi.org/10.3390/suschem2010011

Open Access This chapter is licensed under the terms of the Creative Commons Attribution 4.0 International License (http://creativecommons.org/licenses/by/4.0/), which permits use, sharing, adaptation, distribution and reproduction in any medium or format, as long as you give appropriate credit to the original author(s) and the source, provide a link to the Creative Commons license and indicate if changes were made.

The images or other third party material in this chapter are included in the chapter's Creative Commons license, unless indicated otherwise in a credit line to the material. If material is not included in the chapter's Creative Commons license and your intended use is not permitted by statutory regulation or exceeds the permitted use, you will need to obtain permission directly from the copyright holder.

7

Sustainability in Manufacturing Transforming
Envisioning the Factory of the Future

Nico Hanenkamp and Oliver Zipse

7.1 Introduction

Sustainable production has been the focus of researchers and practitioners for more than two decades. In the beginning, the research largely addressed aspects such as increasing resource efficiency or avoiding hazardous materials in isolation; however, a common understanding exists between academia and industry that sustainability covers a broad range of economic, ecological, and social aspects. This approach is also reflected in the 12th goal of the sustainable development goals (SDG), which is "responsible production and consumption" (UN General Assembly, 2015). Today, the scarcity of material or human resources and increasing environmental and social regulations mean that manufacturing companies must not only address individual aspects of sustainability, but they must also develop an overall strategy and concept for their implementation. This chapter examines how companies can implement this ambition within their own existing manufacturing processes.

As discussed in the previous two chapters, achieving the goal of responsible production requires a new, circular approach to product design (Chap. 5) that has far-reaching implications for sustainable value chains (Chap. 6). Before

---

N. Hanenkamp (✉)
FAU Erlangen-Nürnberg, Fürth, Germany
e-mail: nico.hanenkamp@fau.de

O. Zipse
BMW AG, Munich, Germany

the next chapter (Chap. 8) discusses the technological disruptions that can drive the transition to climate-friendly mobility, this chapter looks at *Sustainability in Manufacturing* as a critical step in this transition journey. While the design of products and value networks is vital, it is through the manufacturing process itself that the involved companies can directly modify their material, energy, social, and environmental footprints.

The purpose of this chapter is to discuss the contributions, tools, and challenges of using sustainable manufacturing to advance the goal of responsible production. The chapter is divided into three parts. Section 7.2 begins with a brief overview of the origin and definition of sustainable manufacturing and then launches an explanation of the three dimensions of sustainability and their implications for manufacturing. The presentation of three use cases illustrates how sustainability is managed at the operational level. Finally, future research perspectives regarding energy use, manufacturing technologies, and circular processes are discussed. Section 7.3 presents the expert conversation between Prof. Oliver Zipse, Chairman of the Board of Management of Bayerische Motoren Werke (BMW) AG, and Prof. Dr.-Ing. Hanenkamp, Institute of Resource and Energy Efficient Production Machines at FAU Erlangen-Nürnberg. Section 7.4 shifts the focus to the sustainable factory of the future, and the chapter concludes in Sect. 7.5 with a short summary and a link to Chap. 8 on *The Power of Technological Innovation*.

7.2 The Three Dimensions of Sustainable Production

Even after almost three decades of research and practical implementation, no common definition exists for sustainable manufacturing (Moldavska & Welo, 2017). However, a consensus has been reached that sustainable manufacturing must cover the three dimensions of economic, ecological, and social aspects (Von Hauff & Jörg, 2017). Although the lack of an abstract definition may seem unimportant at first glance, researchers claim that its absence creates challenges when attempting to take sustainability concepts from theory to practice in the production environment and on the shop floor. Whether sustainable manufacturing is an environmental initiative, a systematic process, a paradigm, or a balance between the dimensions also remains in question. Since the 1990s, a variety of definitions have emerged, but these have served to create more confusion than clarification. The U.S. Department of Commerce defined sustainable manufacturing in 2008 as "the creation of manufactured products that use processes that minimize environmental impacts, conserve energy and natural resources, are safe for employees, communities, and consumers, and are economically sound" (cited in Haapala et al., 2013, p. 041013–2). Since then, research and practice have either referred directly to this definition or adopted similar terms.

The ecological dimension is directly impacted by manufacturing due to the use of (non)renewable resources and the release of emissions into the environment. While the use of renewable resources must not exceed the rate of regeneration, nonrenewable resources should only be used if the possibility of substituting them exists in the long term. From the point of view of an individual company, the economic dimension means reducing the life cycle costs of equipment and manufacturing costs. Finally, the social dimension addresses the needs of employees and society in the manufacturing environment and supply chain. It covers both the health and safety requirements within the production and targets equality among employees with diverse backgrounds while also addressing social aspects within the supply chain (human rights, working conditions, etc.). In the past, many companies prioritized economic and environmental aspects in their sustainability strategies; however, the upcoming demographic change to an aging population in developed countries, which limits the availability of human labor, is now forcing the manufacturing sector to put more emphasis on social aspects (Yuan et al., 2012). Finally, research has shown that the dimensions of sustainability are strongly interlinked, so the full potential of sustainable manufacturing can only be realized by consistently adopting a three-dimensional (3D) approach (Stark et al., 2014). Upcoming regulations, such as the European Sustainability Reporting Standards (ESRS), with their defined structure of reporting elements and key performance indicators (KPIs), can guide practitioners during implementation (European Financial Reporting Advisory Group, 2022). The combination of ecological, economic, and social aspects simultaneously increases a company's competitiveness, as reflected in improved business performance for companies with a consistent three-dimensional approach to sustainable manufacturing.

Manufacturing companies have always striven to improve their operational performance and have developed appropriate principles and management systems, such as lean management, green manufacturing, or Six Sigma. These mature systems already contribute to sustainability in production; however, practices such as lean management alone are insufficient to address all sustainability aspects (Hartini & Ciptomulyonob, 2015). One reason is that the different types of waste only partially address sustainability aspects and do not necessarily focus on a life cycle perspective. Therefore, the challenge from an implementation point of view is to integrate different concepts and management systems, each with a specific focus and expertise, to provide overall sustainability to manufacturing.

The typical research objects tackled with regard to sustainable manufacturing include technologies, the product life cycle from a holistic perspective, value-added networks, and the global manufacturing impact. For each group of research objects, the three dimensions need to be addressed equally.

7.2.1 Practical Perspectives on Sustainable Manufacturing

The following section illustrates the successful implementation of sustainable manufacturing by comparing *three use cases* from BMW's iFACTORY, each with an equal focus on each of the three dimensions but covering the different groups of research objects. With the iFACTORY, BMW addresses the three pillars—LEAN, GREEN, and DIGITAL—thereby setting the direction for the transformation of manufacturing expertise throughout the entire production network (see BMW AG, 2022). This means:

- LEAN—efficient, high-precision, and flexible,
- GREEN—Resource-optimized and circular
- DIGITAL—A new level of data consistency through the efficient use of AI, data science, and virtualization

The first use case shows that incorporating innovative circular materials and systems helps to conserve resources and creates ergonomic benefits for associates. To conserve even more resources, the BMW Group has implemented various projects in packaging logistics. These aim to reduce carbon dioxide (CO_2) emissions in cooperation with suppliers and to implement the principles of circular economy to the greatest extent possible. European plants are increasingly using recycled materials for packaging. In 2022, new contracts for reusable packaging in logistics specified almost double the quota of recycled material, increasing from approximately 20% to over 35%. CO_2 emissions are also being reduced through the use of alternative sustainable materials, less single-use packaging, lightweight packaging, and reduced transport volumes. The BMW Group plans and monitors the effects of individual measures via a CO_2 calculator for packaging.

A second example of innovative production processes with positive reductions in energy and water consumption is the so-called dry scrubber. In a major step toward greater sustainability, paint shops no longer wash away excess paint particles with wet scrubbing but instead are switching to a system of dry separation. In the spray booth, any overspray that does not land on the car body is now collected using limestone powder rather than water, thereby considerably reducing water consumption. Another major advantage is that, unlike wet scrubbing, dry separation can be carried out in up to 90% recirculated air. This means that only 10%, rather than 100%, of the air has to be brought up to the required temperature and humidity, thereby saving vast amounts of energy. The limestone powder also does not need to be processed and disposed of, unlike contaminated water. Instead, it can be returned to the material cycle—for use in the cement industry, for example.

The third use case pays in directly to all three dimensions of sustainable production. A 3D human simulation introduces a virtual model of a human into a virtual production environment. It uses a combination of connected planning data to simulate the complete production and assembly process in 3D. Through this, valuable information can be gathered by simple means, such as planned time analysis, ergonomics assessments, workplace optimization, and validation of planning. This enables optimization of process engineering, the conditions for production workers, and process maturity right at the start of production.

7.2.2 Research Perspectives on Sustainable Manufacturing

Sustainable manufacturing offers a broad spectrum of research opportunities. Due to the interdisciplinary character of sustainability studies, research on the social, economic, and ecological dimensions requires different research competencies. Because of this complexity, this section focuses primarily on the engineering perspectives involving energy, circular processes, and manufacturing technologies and strategies.

With regard to *energy* in the context of sustainable manufacturing, four main research perspectives can be identified. Improving energy efficiency has long been a major focus of research and practice in the past. In addition to energy efficiency (i.e., the relationship between the value created and the energy used; DIN, 2011), energy flexibility requires consideration in the future (Popp, 2020). Energy flexibility describes the ability of a factory or a process to adapt to a volatile energy supply with no negative effects on productivity, quality, or delivery service (VDI, 2020). Overall, 16 flexibility measures have been identified that can be assigned to the factory, production, or process levels. From a research perspective, manufacturing processes, operations management practices, and digitalization technologies all need to evolve to address both energy flexibility and efficiency.

The second perspective involves the substitution of fossil energy sources with renewable energy sources and technologies within a factory. Currently, a strong trend is evident toward the electrification of industrial processes (Wei et al., 2019). With the decreasing price level of solar panels and increasing battery storage capacity, the integration of volatile energy sources to operate industrial processes with a continuous demand is becoming both feasible and advantageous. Although industrial processes cover a wide range of temperatures, electric heating systems, high-temperature heat pumps, or solar thermal technologies can easily generate lower temperatures up to 140 °C.

The third perspective focuses on the systematic change observed across the entire energy supply chain for electricity, from generation to consumption. Decentralized energy generation using photovoltaic systems can now partially replace the traditional external energy supply generated by large power plants and transported over long distances. These approaches can help reduce costs and increase energy resilience.

Finally, production systems and factories based on direct current represent a major new area of research. These systems allow an easier integration of renewable energy sources, such as photovoltaics, while also eliminating the need for frequency inverters that lead to efficiency losses, such as harmonics, and enabling an easier recuperation of electrical energy (Sauer, 2020). This broad scope of the entire system of energy supply, transport, and consumption reveals tremendous improvement potential for energy efficiency, flexibility, and substitution.

With regard to *circular processes*, the second area of research in sustainable manufacturing places a strong emphasis on material flows and digitalization. The linear manufacturing approach of "take–make–use–dispose" not only exceeds the waste-carrying capacity of the earth, but has significantly increased the rate of resource extraction in the recent past. In the EU-28, the manufacturing sector generated 10.3% of all waste, making it the third largest contributor after construction and mining (Rashid et al., 2020). Decoupling resource consumption and waste generation from economic growth will require the application of circular manufacturing. The aim of conventional circular or closed-loop systems is to minimize energy and resource inputs, maximize the value generated, and reduce waste and emissions (Nasr & Thurston, 2006). Closing the loop between output and (re)input can be achieved through reuse, remanufacturing, or recycling. In many cases, this approach is limited because the present-day processes and products were not intentionally designed for closed-loop systems, and the effort to implement circularity exceeds the potential benefits.

According to Rashid et al. (2020) and in line with the circular economy definition of the Ellen MacArthur Foundation (2013), a circular manufacturing system is "a system that is designed intentionally for closing the loop of components or products, preferably in their original form, through multiple life cycles" (Rashid et al., 2020, p. 355). Circular manufacturing can operate at the macro-level (e.g., region and smart city), the meso-level (e.g., industrial parks and factory), or the micro-level (e.g., products and processes) (Urbinati et al., 2020). The micro-level is characterized by the shortest loops and thus has the greatest potential environmental benefits. Based on the original 3R concept (reduce, reuse, and recycle), the 6R framework for implementing circular manufacturing systems, which covers the entire product life cycle (reduce, reuse, recycle, recover, redesign, and remanufacture), represents the state of the art for research and practice (Jawahir & Bradley, 2016).

The first R (reduce) refers to the reduction of resource usage in the pre-manufacturing phase, the reduction of energy and material consumption in the manufacturing phase, and the minimization of emissions in the use phase. The second R (reuse) refers to the multiple life cycles of the original product or its components after each end of life (EOL). The third R (recycle) converts material that would normally be considered waste into new material and process input. To gather the product after the use phase, the fourth R (recover) has the task of recovering the products after their EOL. The fifth R (redesign) incorporates products or components from previous life cycles into the next design concept, while the final R (remanufacture) aims to restore used products to their original state. The 6R system combines traditional methods or tools, such as those for energy efficiency, with innovative remanufacturing processes and facilitates stepwise implementation (Brunoe et al., 2019).

Although circular manufacturing offers tremendous potential for sustainability, its implementation is often hindered by heterogeneous barriers. Because different stakeholders are involved, typically including at least suppliers, the manufacturer, users, and remanufacturing experts, the sharing of data and information is a major challenge. Digital twins of material flows can be used to provide and manage complex and heterogeneous data in discrete manufacturing between them (Acerbi et al., 2022). As an alternative to hierarchical data models, blockchain technology has been implemented to share data among different stakeholders (Govindan, 2022). In doing so, these data models describe the relationships between processes and material flows, reveal optimization potential for circular manufacturing, and deliver consistent and trustworthy data. Thus, in addition to the 6R methodology, the sharing of data and information is considered a prerequisite for implementing circular manufacturing.

Finally, with regard to sustainability in operations, *manufacturing technologies and strategies* represent a third area of research. On the one hand, innovative processes, such as additive manufacturing (AM) or digitalization technologies, have a strong impact on well-established process chains. On the other hand, further development is required to bring innovative technologies to similar quality levels and process capabilities or to scale them up for manufacturing in batch sizes of single products and high-volume production. On the technological side, additive manufacturing (AM) is a primary area of research. For production scenarios with high complexity and low volumes, AM has already demonstrated competitiveness compared with subtractive or formative technologies (Pereira et al., 2019). Due to the reduction in resource consumption and waste generation, AM has a strong positive impact on sustainability. The main challenge for future AM processes and machines is their integration into complete supply chains that meet the requirements of high complexity and large volumes. Other technological challenges arise during the production of electric cars, particularly battery production, or the production of components for hydrogen applications. Both of these examples require innovative, isolated process steps, as well as completely new entire production systems and machines; consequently, low quality levels with high fluctuations are a major concern and have a negative impact on overall equipment effectiveness (OEE) (Schnell & Reinhart, 2016). Finally, process chains for innovative applications or AM will not replace traditional technologies. Further potential for improvement lies in the adoption of hybrid manufacturing approaches, such as configuring the most suitable manufacturing technology for a best practice process chain or even combining technologies with the machine tool (Merklein et al., 2016).

Digitalization and the use of artificial intelligence offer future research perspectives regarding sustainability. At present, Industry 4.0 approaches have been used primarily to address the environmental dimension, but researchers have already outlined research agendas to address the social and economic dimensions in a holistic approach (Machado et al., 2020; Stock & Seliger, 2016). Digitalization techniques, such as the Internet-of-things (IoT) or cloud manufacturing, represent technological tools that must be adopted to pursue sustainability objectives. Artificial intelligence (AI) can be used to manage the complexity of sustainability-related data (e.g., with big data analytics approaches). In any case, digitalization and AI require access to reliable data at the process level.

Manufacturing strategies are an additional area of research. Due to the cross-dimensional nature of sustainability, its strategy must be strongly linked to functional strategies, such as product or process development. Sustainable manufacturing involves technological aspects as well as methods and tools; therefore, a challenge for future research is to integrate well-established management processes, such as quality and supply chain management, and production systems, such as lean management, with sustainability approaches. Replacing existing processes and tools is not recommended; rather, these should be further developed by considering sustainability aspects (Pampanelli et al., 2014).

In summary, various aspects of future research on energy, circular processes, and manufacturing technologies have been highlighted, without claiming to be exhaustive. An important point to note is that intrinsically motivated employees drive the transformation to sustainable manufacturing. They use valid and real-time data in their decision-making to achieve specific and individual sustainability goals. Therefore, in addition to the technical and organizational challenges described above, a suitable qualification concept is of particular importance. To achieve broad acceptance for the implementation of sustainable manufacturing, specific training content and programs with theoretical and practical content must be developed for all hierarchical levels within a company.

7.3 Expert Conversation on Sustainability in Production

What Is Important in Managing Change Toward Sustainability?

Hanenkamp: Change management is an integral part of any successful business. Sustainability brings with it a whole new set of challenges and thus changes. How would you describe the strategic approach to managing change toward sustainability? How do you manage conflicts related to sustainability?

Zipse: Change management is necessary, especially in a high-investment industry. Behind us is a big factory. That is a big investment, an investment of about 2 billion euros. There are certainly good arguments not to change anything about that. So, we need a method to develop a corporate strategy that also takes into account external inputs and answers the question: Is the status quo—including the innovation structure, customer behavior, and cost structure—sustainable in the future? It is necessary to question this status quo at any time. If the status quo of your methods and processes, as well as of your corporate culture, is not good enough, you must change. At

BMW, we coined a term to describe our desired culture. We call it "Be more BMW." Everyone at BMW knows what BMW should be: entrepreneurial, highly innovative, and building the best cars in the world—the ultimate driving machines. At the same time, however, this term stands for a sustainable and profit-oriented strategy. There are diverse requirements, but everyone at BMW knows that this is a solvable equation.

Hanenkamp: Sounds like a continuous journey.

Zipse: It really is. To achieve that, you have to change every day. You have to look for better opportunities every minute, and you have to disregard the status quo if it is not good enough for the future of the company. In production, we all know the old principle of Kaizen (continuous improvement), where all employees consciously question their own activities again and again and constantly improve the way they work. Change management is about looking not only for the big, visible steps, but also for the small, everyday improvements. If you reduce your energy consumption by 30% this week, why not add another percent or two the next week? And it never stops. Change never stops, and there is never a best possible process. Manufacturing is made up of thousands of processes, so it is extremely important that this optimization process never stops. It is a cultural issue but, of course, it is also a technical issue.

Hanenkamp: I understand that continuous improvement is an integral part of a successful company and is also essential for sustainability. We have many processes in place for continuous improvement: quality management systems, lean management culture, etc. There are overlaps, for sure. How do you plan to implement sustainability management in the future? Will there be a standard, a separate sustainability toolbox, with all the sustainability methods? Or will we find a way to integrate these aspects into other management processes?

Zipse: Integration is key. You cannot say, "On this side of the room, we do sustainability. On the other side of the room, we leave it as it was." It is an integrated approach. In production, especially, sustainability comes in two steps. The first and the best thing is that you do not use resources at all. You simply minimize the use of resources in the sense of resource efficiency. Use less light, less energy, and less material, the traditional Kaizen way. This becomes critical because the energy that is not used is the best thing for sustainability. The second step is technical and deals with the question: What kind of new processes can you implement to help you achieve your sustainability goals? What is the role of the digital arena in improving your processes? What kinds of new technologies can you use to be more sustainable? So, in manufacturing, we have two frameworks to be sustainable: resource reduction and technological advancement.

After Kaizen and continuous improvement, what do you see in academia as the next step in optimization? Do you see anything that will dominate the next 20 years of production? Specifically, sustainable manufacturing?

Hanenkamp: First, we need to integrate sustainability aspects into our existing processes and culture. Second, we need to open up to sustainability, as well as to digitalization, and improve our ability to create a digital twin of all production processes and steps. But many open questions remain. We have to figure out how to do this systematically: how to collect data, structure it, make it accessible over time, and maintain it properly. This is our task for the future: to integrate the knowledge and experience that we have from several decades since the early days of Kaizen culture and quality management systems and mix it with digital opportunities. We need to address our processes, first and foremost, without forgetting the corporate culture and mindset of our people.

Is Recuperation a Promising Technology for Sustainable Energy Production Systems?

Zipse: In our factories, we are used to running all our machinery and tools on alternating current (AC). The iX runs on direct current (DC), which is why we can recuperate. When the car brakes, we recuperate the kinetic energy of the car. If you look at a factory, everything is moving, and, of course, everything needs to be accelerated and decelerated. If we had a direct current plant, we could use all that recuperating energy and put it back into the system. We've identified this as an important area of research, and we are very close to some applications.

Is this something that could be a next step in a sustainable, energy-efficient production system?

Hanenkamp: Yes, for sure. This is a very important aspect. There are many other aspects that you can integrate, such as bidirectional loading. What we have to understand is that direct current is more efficient in terms of transferring energy from supply to demand because of harmonics losses. There is tremendous potential in avoiding these losses. The benefit is that the production machines do not necessarily have to change, but we have to reconfigure, redesign the energy supply structure within the plants, and integrate DC principles, and then, the potential is huge.

Zipse: We are thinking along similar lines. It is about questioning how we have thought about energy in the past.

Hanenkamp: Absolutely. From an energy point of view, our whole mindset has to change. In the past, we looked at energy as an unlimited resource; therefore, we did not think much about it. But now, if we look beyond the direct current that could come from renewable energy systems, we see that in many manufacturing plants we have more distributed energy generation systems—thermal block-type power plants, renewable energy systems, etc.—which means that our supply varies over time. We also need to integrate storage systems. In the past, we spent a lot of time and effort trying to find a single stable operating point for the plant. Today, the challenge is to find several of them, because we have to constantly adapt to this fluctuating supply. This is a great opportunity for the future.

What Potential Does System Coupling Hold?

Zipse: You mentioned the topic of system coupling, which is critical for a plant like this, as we have a lot of energy subsystems. Often, the output of one energy system can be the input of another one. For example, the heat we generate in a power-heat coupling can be used in our paint shop. Combining these different systems has an enormous effect. Another example is one we introduced more than 10 years ago: A new paint we introduced made the so-called wet process obsolete. Just by eliminating this one process, we were able to reduce energy consumption by 30%. This phenomenon of looking into product and process design together in terms of sustainability is very common today. Look at the bionic design systems using additive manufacturing technologies. We have brought the cost down—they are still too expensive to be scalable—but every year we take another step. Then, you have product design, weight reduction, and resource efficiency, all in one. If you look at product design and production design pulling together, there is still undiscovered potential.

Hanenkamp: Talking about system coupling, I completely agree. There is huge potential that we can tap into. Production facilities and the technical building infrastructure are often not really coordinated. But it can be done. The technical building infrastructure sometimes consumes up to 50% of the energy of a plant. It runs completely independently of what happens on the shop floor. So why must we turn on the heat 5 min before the shift ends? It does not make sense. Sometimes these processes just have very simple controls like minimum/maximum temperatures. If we could find a way to have something like a projection and see what is coming up, then we could easily adjust the control parameters and not have to turn the heaters off 5 min before the end of the shift. It saves a lot of energy, and it is very easy to do. Today, we can access control parameters via standardized interfaces, but we have to model our process and our production and do a projection. From there, we can access this potential. There is no need to couple production systems at the machine level—a lot of that has already been done. However, the bigger potential is the coupling of the technical building infrastructure with the shop floor.

What Role Does AI Play?

Zipse: What is the potential for the use and implementation of AI in production?
Hanenkamp: There is potential, but it is not an easy thing to implement. The challenge we have to solve is not just to implement islands here and there. AI systems already exist today. We can think of vision systems for quality assurance, for example. We have been doing that for 20 years for specific applications. But the bigger potential is to take a common data perspective to see correlations between two different processes.

Before you can talk about AI, you have to talk about digitalization: You have to have data. It is not just the basis of AI applications, but it is needed to make any kind of fact-based decisions. The challenge that comes with data is that once you invest in data collection and data gathering, you have to do it efficiently. It does not make sense just to collect data, put it in a box, and then figure out what to do with it. You have to allocate it to your specific use cases and what you want to accomplish with it. Otherwise, you overengineer the data. You simply collect it, and you have to manage and store it for a long time. That takes time, money, and energy.

Based on data, we can build AI applications, using data for training. A wide range of AI systems are available—but we need to gain experience regarding which system to use in which application. We have to get over the perception of AI systems as a black-box thing that we do not understand: We throw data in and get data out. We need to have more experience in how to parameterize a neural network system and apply appropriate systems to different use cases.
Zipse: This is a fascinating field of research—and it also reaps the secondary effects of AI.
Hanenkamp: Secondary effects? Can you explain?
Zipse: In the past, we needed perfect lighting for pattern recognition on car surfaces. Pattern recognition was always done with a liquid crystal display (LCD) camera. We would look at it exactly, and if a pattern was not exactly like the perfect condition, we would recognize a quality defect. But with AI, you can train imperfect lighting conditions. In a factory, lighting might differ during the day, during the night, and so on, and AI gives you a lot more flexibility, even during the darker parts of the day, by applying pat-

tern recognition. This means that all the lighting that was extremely energy-intensive could also be spared and energy saved.

That is the secondary effect of using AI. We should think not only about the application in a specific algorithm or a neural network, but also about the secondary effects. AI allows you to be imperfect and much more flexible, which is really exciting potential, especially in the production environment.

How Difficult Is It to Get Buy-In for Change? What Role Do Cooperation and Transparency Play in This Process?

Hanenkamp: What is your experience with the acceptance of sustainability-based changes? When you look at your workforce, are they all open to thinking about sustainability issues? Do some of them see it as a threat?

Zipse: The really amazing thing is that our team at BMW and our employees are very willing to contribute. I have never seen it as a threat. We are looking for new ways to make the company more effective and ultimately more successful, day by day. Sustainability, resource reduction, and improving the quality of our products every day are combined and aligned, not differentiating goals. They are on the same sheet of paper. As society changes, all our employees want to help make processes and products more sustainable and use less energy, because this has become mainstream thinking in our society. If we did not demand a highly sustainable working attitude, our team would be disappointed. In our sustainability strategy, we have the full support of our employees. They like to contribute. Of course, this is also a cultural thing. We want to win this game. The greenest electric car has to come from BMW. That may be easy to say as a goal, but it is not easy to achieve. When we have our integrated report, it will measure who has the lowest resource footprint: energy consumption, CO_2, or all kinds of emissions. We also count on human resources: the tons of labor extracted. All kinds of KPIs you can think of. Then, this will become a field of competition, and BMW is a competitive company.

Hanenkamp: I completely agree. If we go beyond our own organization and look at the supply chain, a car company has a low internal value-add. The majority of the value-add happens upstream in the supply chain. Now, we want our suppliers to contribute to CO_2 reduction as well. Imagine that your in-house processes are already sustainable and set in place. How will you work together with your suppliers to further reduce your carbon footprint? What will this continuous improvement look like outside of your own organization?

Zipse: We have three effects. If you look at the normal production route of an electric car, its CO_2 footprint is higher than that of a combustion-engine car, assuming that 50% of our energy here in Germany is not renewable. This is not sustainable, but we are improving that every day. However, it means that our supply chain has to contribute better results in the production cycle to reduce that footprint. If we take the status quo and ramp up our electric mobility strategy, we will actually increase the carbon footprint in the supply chain. We would be doing exactly the opposite of what we actually want to achieve.

Hanenkamp: So what is the way out?

Zipse: The only way to make sure your suppliers' production is sustainable is to work closely together. For example, if you want your battery supplier to use green energy for cell production, you need to agree on that in your contracts with them. The next step is to have shared transparency on the footprint. That is why we founded the automotive alliance Catena-X, together with SAP, Bosch, ZF, Siemens, Telekom, and so on, to get digital transparency across many companies. This network is growing fast and has strong support from the German government. It is the Internet of companies. We have Industry 4.0 in our own factories as the status quo, and the next step is the Internet of companies, Industry 5.0. A new era where we can document complete supply chains in terms of CO_2 footprint, quality issues, compliance with emission standards, and so on.

Hanenkamp: I completely agree. You cannot improve if you do not have transparency on the baseline. Many suppliers I have talked to are now challenged by this increased transparency, but there is no alternative. To improve, we want to know where to start. We need to collect data efficiently, highlighting waste, losses, and emissions throughout the year. We also have to break it down into smaller reference units, not yearly, monthly, or weekly, but on a daily level, down to a single piece, so we can see deviations over time. We can then see unstable processes precisely and can act on them specifically. This also has a tremendous impact on overall CO_2 emissions. Transparency is first. This is where digital technologies are going to help us gather data, allocate it correctly, and then use it for improvement.

Zipse: It is interesting: The steep drop in the price of sensors—temperature, pressure, and all kinds of IoT sensors—and at the same time the advent of big industrial clouds that are not very expensive and legislation that requires transparency. Together, these three things have an enormous impact because, all of a sudden, you have the tools in your hand to provide transparency. It is no longer a technical issue. You can measure almost any physical state in the supply chain, in a factory, or even in a car. Then, it becomes a matter of collaboration: Who is willing to share that data? We have started to bring these three things together. We do a lot of contracting, but even better than contracting is cooperation.

Hanenkamp: I totally agree. In terms of data collection and sensors, it is much easier than it was a few years ago. The other thing, from a mindset point of view, is: If I am going to capture data, do I have to collect it forever? All the time? Or does not it make sense to capture it temporarily? If you look at a tooling machine that has to manage all kinds of sensors to measure vibrations and temperature, flow rates, and things like that, it costs 40% more than a standard machine. So, that is something you would not want to do everywhere. Measuring vibration, for example, is only important if you have some specific processes. One solution—if we have the sensor and data collecting technology—would be to use it spot-wise and move it from one machine to the next, to be more flexible at a lower cost, yet still get the same information. The other thing is that sometimes we tend to buy a machine that already has all the sensor technology from the supplier. Why do not we do a part of it ourselves? It is not that complicated to put a sensor here and there, but it is still enough to see deviations.

Zipse: I'm glad that you mentioned that. Maintenance is mainly about existing machinery. An existing press shop, for example, lasts 40–50 years. Of course, you have the existing technology, but you can always reequip it with additional sensors. This is actually mainstream: It is not about buying new machinery; it is about digitizing existing machines with sensors. Then, you can improve your maintenance cycles, do preventive maintenance, and see huge effects. These are truly exciting times.

What Role Do Smart Cities Play?

Hanenkamp: Allow me to leave production and jump to smart cities. As I understand it, there is no common definition, but from my understanding, smart cities balance the economic aspects of companies, the people who live there, and other aspects. We balance ecological, economic, and social aspects, as well as digitalization and system coupling. These things we have already discussed as key enablers. If we look ahead, how much will smart cities change the position of manufacturing companies in terms of their locations, how they operate, and the availability of labor, for example?

Zipse: A smart city must be intelligent, as the name implies. It has to reflect the reality of its inhabitants' lives, and of course, it has to be willing to invest, you know. We're sitting here right next to our Munich factory, which is right in the middle of the city. I am a firm believer that there is no contradiction between industrial work and city life. It is possible. A modern city is a synergy of industrial and residential life. It is not the industry that is disappearing from the city. On the contrary, we have this factory here in Munich, and it is very much integrated into its community here, providing jobs and how people get here. We spend a lot on people, on public transport, on company bicycles, and so on.

This is my idea of a smart city: It must be intelligent in terms of providing the right kind of transportation for people, from bikes to cars to buses and public transport. It is a combination of all of that, and it has to have a government structure that is willing and ready to invest. This is because mobility, in particular, depends very much on where the intelligence of the individual mobility lies: Is it in the car, or is it in the city itself? In different parts of the world, really smart cities are developing in which all the intelligence is put into the infrastructure of the city. Then, it does not have to be in the car. Smart cities are about combining individual mobility needs, the intelligence of the city's infrastructure city, and the reality of the people who live there.

7.4 The Sustainable Factory of the Future

Increasing resource efficiency and implementing sustainability are key challenges for industry in the future. In 2010, industrial production was responsible for more than 30% of global greenhouse gas emissions, which is only one environmental factor. Companies are therefore called upon to make a significant contribution to reducing their environmental impact. The vision of sustainable production goes beyond the isolated ecological dimension and takes into account social responsibility, competitiveness, and environmental protection. Furthermore, positive interactions between the three dimensions lead to additional benefits for all stakeholders (Stark et al., 2014). The implementation of sustainable production is currently driven by both economic incentives and regulatory requirements. Successes have already been achieved, such as increased energy efficiency and the positive effects of introducing sustainability management systems.

On the one hand, the challenge is that no universal blueprint has been drawn for the transformation to sustainable manufacturing, so the journey must be planned, implemented, and tracked individually. On the other hand, the majority of manufacturing companies rely on experience to manage complex changes, such as the transformation toward a lean company or to implement Industry 4.0 principles and technologies. Despite these challenges, new technologies, such as artificial intelligence, or accepted standards, such as the life cycle assessment (LCA) methodology, can be applied to improve environmental impacts. In most cases, the transformation must follow a brownfield approach; that is, the existing equipment and infrastructure have to be upgraded and integrated into the new production system in combination with new production processes, such as additive manufacturing. In summary, this change is a complex transformation, the key principles of which are discussed in this article from an operations point of view. First, sustainable process and factory planning, as well as operations management, will be presented. Second, the contribution of digitalization and artificial intelligence in an industrial context is considered. Third, the impact of sustainable manufacturing standards and methods will be highlighted. Finally, the coupling of direct and indirect manufacturing systems and the integration of urban production in smart cities will be discussed.

7.4.1 Sustainable Manufacturing Processes along the Life Cycle

Research and practice agree that sustainability aspects can only be addressed if improvements consider all phases of the product life cycle (i.e., the development phase, the manufacturing phase, the use phase, and the end-of-life phase). Interactions between the phases need to be considered (Liu et al., 2019). In the following section, challenges and opportunities along the design and manufacturing life cycle phases that impact sustainability will be discussed.

The design of a manufacturing system is critical because changes in later stages can only be implemented with a substantial effort. Moreover, decisions have to be made under uncertainty and undefined boundary conditions. The dimensioning of a production system, and especially its capacity, is closely related to its sustainability impact and must therefore be derived using a systematic process. For example, if the technological and production capacity after ramp-up significantly exceeds the current process demand, this effect leads to inefficient operating points in the manufacturing phase. In addition to sizing, the specification of the production equipment in terms of its process steps and manufacturing technology is of great importance. Value-adding steps must be optimized, while non-value-adding steps and process waste must be minimized. If we are to manage the sizing uncertainty and define optimal processes, we must have reliable process data. However, collecting process data based on physical testing and design of experiments (DoE) is not only time-consuming; it is also often impossible to obtain because the manufacturing equipment is not yet available at this early stage. As digital twins and simulation technologies provide virtual representations of systems along the life cycle, they can also be used to model the dynamic behavior of the production system in terms of sustainability (Negria et al., 2017). Input factors, such as raw materials, consumables, and energy, and output factors, such as productivity or waste streams, can be determined based on varying operating conditions without performing physical tests. To minimize the implementation effort and to achieve high accuracy of digital twin modeling, most practitioners and researchers follow a systems engineering approach (i.e., the production system is broken down into smaller units for which reliable digital twins are developed based on existing data; Computer-Aided Design [CAD], Product Lifecycle Management [PLM], etc.). With increasing maturity and given the physical availability of manufacturing equipment and process design, congruence between digital twins and physical systems must be achieved. Finally, the increasing application of digital twins not only increases the efficiency in advanced product quality processes (APQP) within a single company, but they can also be used to model interdependencies related to sustainability at the interorganizational level.

During the ramp-up and in series production, the focus must be on efficiency. One key metric is overall equipment effectiveness (OEE), with its three components: loss of availability, loss of performance, and loss of quality (Focke & Steinbeck, 2018). Since losses of availability include all downtime of the manufacturing system, the indicator shows the percentage of a period that the system is in stable operation. Sustainability is negatively impacted because a high level of availability losses requires additional capacity reserves to meet the total demand, with a negative impact on space, and frequent interruptions to operations result in ramp-up losses of energy, personnel, and raw materials. The second component of OEE is performance loss, which describes whether the production system is at its optimal operating point regarding energy consumption, material input and output, and personnel. Temporary or permanent deviations require additional production capacity. Finally, quality loss is the amount of scrap and rework that occurs in the process chain. Poor quality levels have a direct impact on sustainability because the initial raw material is not processed into finished goods. Emissions, material consumption, etc. from the raw material generation phase have already been incurred but cannot be used to create products or added value. To incorporate sustainability aspects, a stronger focus is needed on inputs and outputs that have not been considered before, such as emissions and waste streams. These extensions to existing key performance indicators will provide reliable and consistent data for effective sustainability decision-making. Practice and research reflect that modern shop floor management systems follow this approach and include sustainability metrics. Based on this information, anomalies from the defined operating points can be identified, and appropriate countermeasures can be initiated on the shop floor (Cerdas et al., 2017).

For the production system design and manufacturing phases, efficiency is the central objective. Industrial production processes transform material, energy, and other inputs into finished goods, delivering added value as well as by-products, such as waste streams and energy losses. To minimize these by-products, circular production processes must be developed and installed (Gupta et al., 2021). The concept of the ultra-efficient factory relies on reuse of all types of waste and energy in two main energy and material recycling loops. In the first loop, wasted energy and materials are fed directly back into the manufacturing phase. In the second loop, the product is returned to the supply chain at the end of its useful life. This minimizes downcycling of material (i.e., the use of material for lower performance applications). This means that sustainability is based on efficient product generation processes and on efficient recycling and remanufacturing concepts that must be designed into the manufacturing design phase.

7.4.2 Digitalization, Artificial Intelligence, and IoT

Digitization, Industry 4.0/IoT, and artificial intelligence have significant potential to allow manufacturing companies to implement sustainability (Stock & Seliger, 2016). However, an important point to consider is that digitization is not an end in itself, and its implementation requires a systematic approach. A generic model is the manufacturing analytics approach with four levels: (1) visibility, (2) transparency, (3) forecasting ability, and (4) prescription. This approach has been developed for the systematic implementation of digitization technologies (Meister et al., 2019). It ranges from lower levels of digitalization, such as simple data collection, to the modeling of complex system behavior using artificial intelligence. Although it is not primarily intended for the implementation of sustainability, it represents a systematic approach to the acquisition, handling, and management of data for specific objectives. When applied to sustainability issues, this analytics approach can be used to better understand correlations, optimize processes, and anticipate and prevent negative impacts on the three dimensions of sustainability.

The objective of first-level visibility is to capture data from the shop floor. Accessing shop floor data is a hurdle because it either has to be accessed through a wide variety of different protocols (OPC Unified Architecture [OPC UA], MTConnect, EuroMap77, etc.) or is not accessible at all. Since sustainability data are not always part of existing protocols, retrofitting existing machines with Internet-of-things–compatible sensors is often necessary. The result is that the aggregated data at each time step are stored on a common, often cloud-based platform. The objective of transparency (2) is to systematically identify the root causes of specific problems and deviations, such as the increased use of energy or material consumption. Individual and specific KPIs, for example, for different functional units, can be extracted. The forecasting ability (3) enables us to make projections of trends in the future and to proactively manage deviations. This ability can be used, for example, for the demand-side management of production equipment and the ramp-down of lower priority processes and machines in the event of energy shortages or price increases. At the prescription level, courses of action are being proposed.

As described above, the analytics approach and the application of artificial intelligence methods and tools are highly interdependent. AI algorithms require consistent data on a continuous basis, which is provided by the four-stage model (Weber et al., 2019). When this condition is met, AI methods can first be applied to reduce the complexity of data lakes. For example, principal component analysis can be used to identify the primary drivers of sustainability improvement actions. In addition, black-box AI systems, such as neural networks, can be applied to speed up simulation runs of energy consumption under different or uncertain conditions.

7.4.3 Application of Sustainability Standards

The development of methods and standards for sustainable production has long been a focus of research and practice. The goal is to make visible the relationships between production, consumption, and disposal and to assess the impacts of economic activities. Life cycle assessment (LCA) has emerged as the most important and accepted method from a technical perspective (Hagen et al., 2020). It is embedded in the ISO 14000 series of environmental standards that address environmental management issues associated with production processes and services. Common to all sustainability standards is the breaking down of industrial value streams into process modules for which mass flows (raw materials and fuel inputs, products, by-products, and waste), energy inputs, and emissions to water, air, and soil are analyzed. While the data of Scopes 1 and 2 of DIN EN ISO 14064 can be collected internally, cooperation with suppliers is required to collect data for Scope 3 raw and operating material inputs. Scope 3 CO_2 emissions are particularly relevant, as they can account for up to 50% of the total footprint (Gross & Hanenkamp,

2021). In practice, suppliers are under increasing pressure from their customers to provide data on CO_2 emissions.

Environmental impact categories are assigned to the life cycle inventory analysis, and their quantification allows us to focus on prioritized environmental impacts. In practice, the application of sustainability standards with precise data requires a high level of technical effort due to its complexity, as well as extensive methodological knowledge and expertise. As a result, assessments are often conducted on a project-by-project basis and are static in nature, making them unsuitable for the operational optimization of production processes. A dynamization of the LCA (i.e., continuous generation with real-time data) can be used to derive precise measures for the operational optimization of the production processes on the shop floor (Cerdas et al., 2017). Finally, to reduce the burden on all stakeholders in the supply chain, the exchange of sustainability-related data based on trust and using reference data models is required.

7.4.4 System Coupling, Urban Production, and Smart Cities

The need for more efficient and sustainable operations necessitates that we do not develop and optimize production systems independently, but rather consider them as interconnected entities. In the circular concept, energy, material, and waste streams from one process must be considered for secondary use in other processes. This can only be achieved by coupling different entities of the manufacturing system. The concept of system coupling allows the physical flow of materials between subsystems, the recuperation and use of wasted energy, such as electricity or heat, and the exchange of information, such as future demand or the current status. The peripheral components within production systems, such as cooling devices, are typically operated using simple control strategies with few set points; consequently, energy demand peaks cannot be avoided. Heating, ventilation, and air-conditioning systems (HVAC) rely on more complex control strategies, but their control parameters are not adapted to upcoming heating or ventilation demands. The prerequisite for system coupling within the factory is the exchange of data between manufacturing systems and technical building equipment. This allows for the identification of optimal operating points that lead to the adjusted control parameters of the subsystems.

Beyond the internal system coupling within the factory boundaries, the manufacturing site also interacts with the local urban environment.

Historically, manufacturing and urban spaces have coexisted, and negative impacts have led to the location of factories on the outskirts of cities. Urbanization, as a megatrend, forces the development of new concepts, such as urban manufacturing or the integration of manufacturing in smart cities (Matt et al., 2020). By definition, an urban factory is not only a factory that is simply physically located in an urban environment; it is one that strongly interacts with other urban entities regarding information, material, and energy flows and that relies on the local market and suppliers (Ijassi et al., 2022). In this way, urban production can contribute to the sustainable development goals (SDG) of affordable and clean energy (SDG 7), decent work and growth (SDG 8), industry innovation and infrastructure (SDG 9), sustainable cities (SDG 11), and responsible consumption and production (SDG 12) (Juraschek et al., 2018). Thus, negative impacts, such as emissions, arise, but positive contributions, such as the availability of jobs in urban production scenarios, also occur. Given the global trend of urbanization, smart cities will also play a central role in sustainability. Although the concept of a smart city has no common definition in research and practice, it has a broader scope than manufacturing (Suvarna et al., 2020). It encompasses all entities within the city (including buildings, transportation, energy grids, health care, manufacturing, and commercial services) that need to be connected. This also means that material, information, energy, and people flows need to be considered and optimized in the context of the city ecosystem.

7.4.5 Summary and Outlook

Four different principles of the sustainable factory of the future were discussed in this chapter. First, the planning and operation of manufacturing processes with respect to sustainability were shown. While, in the planning phase, the dimensioning and specification of the production system are crucial, whereas, in the manufacturing phase, the focus has to be on efficiency and abnormality management. In the future, this will require, for example, bringing sustainability aspects to the daily shop floor management level. Second, the availability of real-time process and manufacturing data is critical. Sustainability-related decisions are often highly complex and require a systematic approach to collecting, processing, and proactively applying manufacturing data. Third, the factory of the future can be assessed for sustainability based on accepted standards. Compared with today's static nature of assessments, the assessments will need to be performed more frequently with minimal effort. Finally, the sustainable factory of the future is characterized by system coupling at multiple levels. Within the factory, production systems, peripheral components, and HVAC systems are physically and digitally connected and operated with global optima in mind. Beyond the physical boundaries of the factory, the exchange of material, information, and energy with the urban space must be considered to have a positive impact on sustainability. Given these directions, research and practice are challenged to develop methods and tools for implementing sustainable factories.

7.5 Conclusion

Sustainability principles are widely accepted, but implementing them in manufacturing is a challenge. The three dimensions of sustainability—the social, ecological, and economic aspects—must be equally considered, as innovation and research are essential for sustainability in operations. Successful companies have a clear vision of sustainable manufacturing processes, high digitalization, and the use of artificial intelligence.

So, how can sustainability be integrated into manufacturing?—We would like to highlight five takeaways from this chapter that invite further discussion:

1. The path toward sustainable production is a continuous process and not a single and isolated project. Companies have to use their experience, methods, and tools of continuous improvement from quality or lean management to plan and implement sustainability measures following long-term objectives.
2. Sustainability depends on the availability of and access to data from various sources, such as production machines, information technology (IT) systems, or manual processes. To achieve greater transparency, IoT sensors and industrial clouds can be used to store and analyze the underlying sustainability-relevant data.
3. Improvement processes, such as Kaizen, have to be extended to develop and optimize production processes physically and through the use of digital twins. This approach allows the use and implementation of artificial intelligence for sustainability aspects.
4. While efficiency improvement measures must be implemented in the short term and at existing manufacturing sites, new investments in equipment and infrastructure must include sustainability aspects as important selection criteria.
5. Recuperation of the energy of production processes contributes to further efficiency improvements. This requires linking energy sinks and sources in

the process chain. Similarly, circular processes for material flows should play a major role in industrial engineering.

On the Road to Net Zero, sustainable manufacturing provides companies with the most direct lever to drive decarbonization and other sustainability objectives in industrial value creation. The ability to innovate manufacturing, however, is also crucial for introducing new technologies in the marketplace. In the automotive industry, disruptive technological transformation is needed to replace fossil-fuel combustion engines with drive-train technologies based on renewable energy. For this reason, Chap. 8 now looks at *The Power of Technological Innovation*.

References

Acerbi, F., Sassanelli, C., & Taisch, M. (2022). *A conceptual data model promoting data-driven circular manufacturing.* operations management research, (Vol. 15, p. 838). Springer Verlag.

BMW AG. (2022). Die Produktion von Morgen: BMW iFACTORY. Retrieved June 19, 2023, from https://www.bmwgroup.com/de/news/allgemein/2022/bmw-ifactory.html

Brunoe, T., Andersen, A., & Nielsen, K. (2019). Changeable manufacturing systems supporting circular supply chains. *Procedia CIRP, 81*, 1423–1428.

Cerdas, F., Thiede, S., Juraschek, M., Turetskyy, A., & Hermann, C. (2017). Shop-floor life cycle assessment. *Procedia CIRP, 61*, 393–398.

DIN EN ISO 50001. (2011). *Umweltmanagementsysteme – Anforderungen mit Anleitung zur Anwendung.* Beuth Verlag.

Ellen MacArthur Foundation. (2013). Towards the circular economy Vol. 1: An economic and business rationale for an accelerated transition. Retrieved September 19, 2022, from https://www.ellenmacarthurfoundation.org/assets/downloads/publications/Ellen-MacArthur-Foundation-Towards-the-Circular-Economy-vol.1.pdf

European Financial Reporting Advisory Group. (2022). Sustainability reporting standards interim draft. Retrieved September 19, 2022, from https://www.efrag.org/Activities/2105191406363055/Sustainability-reporting-standards-interim-draft

Focke, M., & Steinbeck, J. (2018). *Steigerung der Anlagenproduktivität durch OEE-management.* Springer Verlag.

Govindan, K. (2022). Tunneling the barriers of blockchain technology in remanufacturing for achieving sustainable development goals: A circular manufacturing perspective. *Business Strategy and the Environment, 31*(8), 3769–3785.

Gross, D., & Hanenkamp, N. (2021). Energy efficiency assessment of cryogenic minimum quantity lubrication cooling for milling operations. *Procedia CIRP, 93*, 523–528.

Gupta, H., Kumarb, A., & Wasan, P. (2021). Industry 4.0, cleaner production and circular economy: An integrative framework for evaluating ethical and sustainable business performance of manufacturing organizations. *Journal of Cleaner Production, 295*, 126253.

Haapala, K. R., Zhao, F., Camelio, J., Sutherland, J. W., Skerlos, S. J., Dornfeld, D. A., Jawahir, I. S., Clarens, A. F., & Rickli, J. L. (2013). A review of engineering research in sustainable manufacturing. *Journal of Manufacturing Science in Engineering, 135*(4), 041013.

Hagen, J., Büth, L., Haupt, J., & Cerdas, F. (2020). Live LCA in learning factories: Real time assessment of product life cycles environmental impacts. *Procedia Manufacturing, 45*, 128–133.

Hartini, S., & Ciptomulyonob, U. (2015). The relationship between lean and sustainable manufacturing on performance: Literature review. *Procedia Manufacturing, 4*, 38–45.

Ijassi, W., Evrard, D., & Zwolinski, P. (2022). Characterizing urban factories by their value chain: A first step towards more sustainability in production. *Procedia CIRP, 105*, 290–295.

Jawahir, I., & Bradley, R. (2016). Technological elements of circular economy and the principles of 6R-based closed-loop material flow in sustainable manufacturing. *Procedia CIRP, 40*, 103–108.

Juraschek, M., Bucherer, M., Schnabel, F., Hoffschroer, H., Vossen, B., Kreuz, F., Thiede, S., & Herrmann, C. (2018). Urban factories and their potential contribution to the sustainable development of cities. *Procedia CIRP, 69*, 72–77.

Liu, Y., Syberfeldt, A., & Strand, M. (2019). Review of simulation-based life cycle assessment in manufacturing industry. *Production & Manufacturing Research, 7*(1), 490–502.

Machado, C., Winroth, M., & Ribeiro da Silva, E. (2020). Sustainable manufacturing in industry 4.0: An emerging research agenda. *International Journal of Production Research, 58*(5), 1462–1484.

Matt, D., Orzes, G., Rauch, E., & Dallasega, P. (2020). Urban production – A socially sustainable factory concept to overcome shortcomings of qualified workers in smart SMEs. *Computers & Industrial Engineering, 139*, 105384.

Meister, M., Beßle, J., Cviko, A., Boing, T., & Metternich, J. (2019). Manufacturing analytics for problem-solving processes in production. *Procedia CIRP, 81*, 1–6.

Merklein, M., Junker, D., Schaub, A., & Neubauer, F. (2016). Hybrid additive manufacturing technologies – An analysis regarding potentials and applications. *Physics Procedia, 83*, 549–559.

Moldavska, A., & Welo, T. (2017). The concept of sustainable manufacturing and its definitions: A content-analysis based literature review. *Journal of Cleaner Production, 166*, 744–755.

Nasr, N., & Thurston, M. (2006). Remanufacturing: A key enabler to sustainable product systems. *Proceedings of LCE*, 15–18. Retrieved September 19, 2022, from: https://www.mech.kuleuven.be/lce2006/key4.pdf

Negria, E., Fumagallia, L., & Macchi, M. (2017). A review of the roles of digital twin in CPS-based production systems. *Procedia Manufacturing, 11*, 939–948.

Pampanelli, A., Found, P., & Bernardes, A. (2014). A lean & green model for a production cell. *Journal of Cleaner Production, 85*, 19–30.

Pereira, T., Kennedy, J., & Potgieter, J. (2019). A comparison of traditional manufacturing vs. additive manufacturing, the best method for the job. *Procedia Manufacturing, 30*, 11–18.

Popp, R. (2020). *Energieflexible, spanende Werkzeugmaschinen – Analyse.* TU München, Utz Verlag.

Rashid, A., Roci, M., & Asif, F. (2020). Circular manufacturing systems. In M. Brandao, D. Lazarevic, & F. Finnveden (Eds.), *Handbook of the circular economy* (pp. 343–357). Elgaronline.

Sauer, A. (2020). *Die Gleichstromfabrik: Energieeffizient.* Carl Hanser Verlag.

Schnell, J., & Reinhart, G. (2016). Quality management for battery production: A quality gate concept. *Procedia CIRP, 57*, 568–573.

Stark, R., Grosser, H., Beckmann-Dobrev, B., & Kind, S. (2014). Advanced technologies in life cycle engineering. *Procedia CIRP, 22*, 3–14.

Stock, T., & Seliger, G. (2016). Opportunities of sustainable manufacturing in industry 4.0. *Procedia CIRP, 40*, 536–541.

Suvarna, M., Büth, L., Hejny, J., Mennenga, M., Li, J., Ng, Y. T., Herrmann, C., & Wang, X. (2020). Smart manufacturing for smart cities—Overview, insights, and future directions. *Advanced Intelligent Systems, 2*(10), 2000043.

UN General Assembly. (2015). Transforming our world: the 2030 Agenda for Sustainable Development, Retrieved July 6, 2023, from https://www.refworld.org/docid/57b6e3e44.html.

Urbinati, A., Rosa, P., Sassanelli, C., Chiaroni, D., & Terzi, S. (2020). Circular business models in the European manufacturing industry: A multiple case study analysis. *Journal of Cleaner Production, 274*, 122964.

VDI. (2020). *Energieflexible Fabrik - Grundlagen.* 5207 Blatt 1,. Beuth Verlag.

Von Hauff, M., & Jörg, A. (2017). *Nachhaltiges Wachstum..* DeGruyter.

Weber, T., Sossenheimer, J., Schäfer, S., Ott, M., & Walther, J. (2019). Machine learning based system identification tool for data-based energy and resource modeling and simulation. *Procedia CIRP, 80*, 683–688.

Wei, M., McMillan, C., & de la Rue du Can, S. (2019). Electrification of industry: Potential, challenges and outlook. *Current Sustainable/Renewable Energy Reports, 6*, 140–148.

Yuan, C., Zhai, Q., & Dornfeld, D. (2012). A three dimensional system approach for environmentally sustainable manufacturing. *CIRP Annals - Manufacturing Technology, 61*, 39–42.

Open Access This chapter is licensed under the terms of the Creative Commons Attribution 4.0 International License (http://creativecommons.org/licenses/by/4.0/), which permits use, sharing, adaptation, distribution and reproduction in any medium or format, as long as you give appropriate credit to the original author(s) and the source, provide a link to the Creative Commons license and indicate if changes were made.

The images or other third party material in this chapter are included in the chapter's Creative Commons license, unless indicated otherwise in a credit line to the material. If material is not included in the chapter's Creative Commons license and your intended use is not permitted by statutory regulation or exceeds the permitted use, you will need to obtain permission directly from the copyright holder.

8

The Power of Technological Innovation
Driving Sustainable Mobility

Jörg Franke, Peter Wasserscheid, Thorsten Ihne, Peter Lamp, Jürgen Guldner, and Oliver Zipse

8.1 Introduction

The rapid decarbonization needed to meet the 1.5 °C target will require disruptive technological change. In general, there are strong interactions between technological innovation and increased sustainability. So, technological progress can be a key to increased sustainability. In parallel, a stronger focus on sustainability goals requires technological innovation. In this context, technological progress presents both opportunities and risks for many market participants. Emerging technologies are always associated with uncertainties from various sources, which means that their potential, likelihood of occurrence, and the timing are often unclear for a long time (Kapoor & Klueter, 2021). As a result, all relevant stakeholders in business, society, and politics are faced with major challenges.

This is especially true for the mobility transition that is currently taking place in almost all relevant markets in the context of climate change and environmental protection. On the Road to Net Zero, this chapter on *The Power of Technological Innovation* addresses the management of uncertainty associated with emerging technologies in the mobility sector. At the heart

---

J. Franke (✉) • T. Ihne • P. Wasserscheid
FAU Erlangen-Nürnberg, Erlangen, Germany
e-mail: joerg.franke@faps.fau.de

P. Lamp • J. Guldner • O. Zipse
BMW AG, Munich, Germany

of this technological transformation is the drive system and its interaction with the associated energy ecosystem. This chapter thus complements the previous chapters on *Creating Sustainable Products* (Chap. 5), *Transforming Value Chains for Sustainability* (Chap. 6), and *Sustainability in Manufacturing* (Chap. 7) by broadening the perspective to include external factors such as infrastructure and energy systems. These aspects and technology are inextricably linked and can only be evaluated together in terms of carbon emissions and ecological footprint. In parallel, the economic balance must also be considered holistically, as this aspect is critical to the success of the transformation. Forecasts vary widely, ranging from scenarios in which fossil fuels continue to play a significant role globally, to scenarios dominated by e-mobility (Zapf et al., 2021).

The purpose of this chapter is to discuss these corresponding factors in more detail. First, Sect. 8.2 presents the advantages and disadvantages of alternative drive systems. This is followed by an explanation of the motivation for the current technological transformation. Section 8.3 presents the expert conversation by Prof. Oliver Zipse, Chairman of the Board of Management of Bayerische Motoren Werke (BMW) AG, Dr. Peter Lamp, General Manager Battery Cell Technology at BMW, and Prof. Dr.-Ing. Jörg Franke, Institute for Factory Automation and Production Systems at FAU Erlangen-Nürnberg, on the future of drive technology from a business perspective. In Sect. 8.4, Prof. Oliver Zipse, Dr. Jürgen Guldner, General Program Manager Hydrogen Technology at BMW, and Prof. Dr. Peter Wasserscheid, Director of the Helmholtz Institute Erlangen-Nürnberg for Renewable Energy and Chair of Chemical Engineering I (Reaction Engineering) at FAU Erlangen-Nürnberg, engage in an expert conversation on the future opportunities of hydrogen as an alternative energy carrier for the automotive industry. Finally, Sect. 8.5 identifies future directions for research and practice to advance the market viability of alternative drivetrains. The focus is set on the energy ecosystem as an enabler for future drive technologies. The chapter concludes in Sect. 8.6 with a brief summary and a link to the concluding chapter (Chap. 9), *The Road to Net Zero and Beyond*.

8.2 An Overview on Alternative Drive Systems

The European automotive industry is undergoing dynamic change (see Fig. 8.1). In the face of the climate catastrophe, emission limits are becoming increasingly stringent, fuel prices are rising, and individual mobility is being

Fig. 8.1 Overview of the competitive environment in the automotive industry

hampered by regulations, competing mobility concepts, and conflicting customer interests. While Western private passenger car markets tend to shrink, new competitors are emerging, especially from China (Kaul et al., 2019). Most importantly, new technologies are arising that are shaking up the automotive market, which has been fairly stable for decades. Autonomous driving promises completely new business models for passenger and freight mobility, software will increasingly dominate over mechanical functions, and, finally, the internal combustion engine (ICE) will eventually be replaced by electric drive systems.

The ICE was the foundation for the triumphant advance of individual mobility in the twentieth century: robust and reliable gasoline and diesel engines powered, at their peak, almost 100 million annually newly produced passenger cars and trucks, as well as tens of millions of motorcycles worldwide (European Environment Agency, 2019; Umweltbundesamt, 2022; Wang, 2021). The enormously high energy density of fossil fuels allowed enormous ranges of up to 1000 km without stopping for refueling, while the persistently unrivaled low energy cost of oil and its seemingly unlimited availability provided the general public with continent-wide freedom of movement.

Over time, ICE-based drivetrains have evolved into highly complex engineering marvels, improving their energy efficiency and significantly reducing their impact on air pollution and climate change. However, ICE-based road transport is still responsible for about approximately 20% of carbon dioxide (CO_2) emissions, contributes to air pollution especially in large cities, burns the precious natural resource of fossil fuels, and perpetuates dependence on

Fig. 8.2 Push and pull factors underpinning the success of electromobility (own illustration based on International Energy Agency (2022, 2023) and Reitz et al. (2020))

politically unstable or unreliable countries (see Fig. 8.2) (International Energy Agency, 2022, 2023; Reitz et al., 2020).

In this context, political regulations, such as the tightening of European emission limits (Euro 7) and the goal of climate neutrality by 2050 (European Green Deal), are increasingly weighing on the market environment. As ICEs have approached the asymptotic branch of their S-curve, where further improvements require disproportionate effort, and have little impact, many automotive manufacturers (OEMs) have already announced to stop the development of new ICEs.

The electric motor is a comparable old propulsion technology. It is completely emission-free, has an unsurpassed efficiency (~95%; the ICE is barely above 30% efficiency), and therefore causes only about a tenth of the losses of internal combustion engines. It can use renewable energies directly and without great effort, and it does not waste valuable fossil resources. The unrivaled characteristics of electric motors offer a much wider speed range with consistently high torque, thereby eliminating the need for complex shifting transmissions (with up to ten gears) and providing a highly dynamic driving experience. The control dynamics of electric drives, which are an order of magnitude faster, allow the vehicle to be steered longitudinally and laterally over wide ranges without braking, and kinetic energy can also be recuperated in the process (e.g., the BMW iX xDrive50 has a 208 kW maximum recuperation power) (Schwarzer, 2019). In addition, the power density of electric motors is significantly higher, their running smoothness is unparalleled due to the rotating drive, and their wear is negligible due to the contactless power transmission (Parizet et al., 2016; Specht, 2020). Based on the technical, ecological, and economic advantages summarized in Fig. 8.2, it is highly likely that at least the majority of land vehicles will be powered by electric drives at some point in the future.

Electric motors are compatible with a wide range of different drive configurations: As a result, electromobility takes many forms, from battery electric vehicles to hybrid concepts and fuel cell applications. Battery electric vehicles (BEV) are increasingly gaining market share. BEVs incorporate a high-voltage battery, enabling electrification of auxiliary units, brake energy recuperation, and plug-in recharging at a wall outlet or charging stations. The major remaining weakness of BEVs is their comparatively short range (currently about 400 to 700 km), as electrochemical batteries allow only about 5% of the gravimetric energy density of gasoline storage systems (0.5 versus 11.4 kWh/kg) (Sartbaeva et al., 2008; van Basshuysen & Schäfer, 2017). From a technical perspective, this disadvantage is exacerbated by the still considerable charging times (around 30 min), since even an electrical charging power of 250 kW corresponds to only about 1% of the power transfer during refueling. With the current state of technology (SoT), these immense differences can only be partially compensated for by the significantly higher efficiency of electric vehicles, which is a factor of three to four for a typical driving profile (e.g., for a WLTP[1] cycle: the BMW i4 eDrive40 Gran Coupé [250 kW]: 16.8 kWh versus BMW M440d xDrive Coupé [250 kW]: 5.7 l_{Diesel}/100 km, 55.9 kWh/100 km).

Battery technology is currently undergoing continuous development to address the range issue. Lithium-ion batteries are currently the automotive standard due to their robustness, high cycle stability, and a high energy density. Significant increases in energy density are currently being achieved, while costs are falling. While the price per kilowatt hour averaged 600 euros/kWh in 2010, it is expected to be approximately 83 euros/kWh in 2025. In addition to specific energy density and costs, other aspects such as shorter charging times, longer lifetimes, improved temperature performance, and higher reliability are also in focus. Environmental compatibility and the supply of critical raw materials are also important aspects. Battery reuse and recycling are already being implemented, and the corresponding capacities are currently being greatly expanded (Blois, 2022). In parallel, alternative battery technologies are being researched such as solid-state batteries, which allow significantly

---

[1] Worldwide Harmonized Light Vehicles Test Procedure (WLTP): Standardized test cycle determined on test rigs under defined laboratory conditions and based on empirically determined real driving data from Asia, Europe, and the United States. The WLTP cycle has been valid in the European Union (EU) since September 2017 (Verband der Automobilindustrie e.V. 2018).

higher specific energy densities. BMW is also involved in the all-solid-state-battery (ASSB) technology.

Because of the range issue, hybrids are still relevant today. Hybrid vehicles have at least two different energy converters and two different energy storage systems, so they include both an ICE and an electric motor, as well as a fuel tank and a battery. Hybrid powertrains can be classified according to the degree of hybridization and the energy flow. Development has started on micro-hybrid electric vehicles (MCHEVs), which enable a start–stop strategy, and mild-hybrid electric vehicles (MHEVs), in which an electric machine supports the ICE for load-point shifting. From an environmental and climate protection perspective, these measures are no longer sufficient. To eliminate local emissions and provide at least temporary zero-emission driving, the degree of hybridization must allow for full electric driving. Full-hybrid electric vehicles (FHEVs) allow all-electric driving over shorter distances, while providing good efficiencies through load-point shifting and recuperation. However, the electric range and performance are limited. Plug-in hybrid electric vehicles (PHEVs) compensate for the disadvantages by allowing the traction battery to be recharged externally. Figure 8.3 provides an overview of the functionalities of different hybridization strategies.

Regardless of the degree of hybridization, the architectures differ. The serial hybrid uses the ICE as a generator and is driven only by electric motors. This means that the ICE is constantly operating at an optimal operating point. Range extenders are a special form of serial hybrids in which a normally

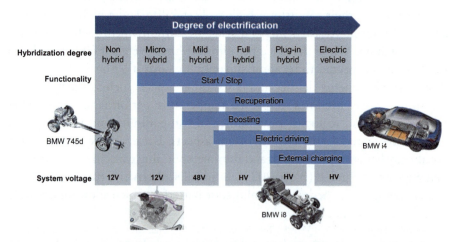

Fig. 8.3 Functionalities of different hybridization strategies (own illustration based on Doppelbauer (2020) and Tschöke et al. (2019))

switched off ICE charges the battery of otherwise all-electric vehicles when needed. The parallel hybrid has a switchable mechanical connection between the ICE and the drive axle, so that either drive can be used for propulsion. These can be dimensioned smaller, accordingly. Mixed forms of serial and parallel hybrids are also available. These power-split hybrids use the power of the ICE for both propulsion and battery charging, resulting in high efficiency over the entire load profile. Although hybrids, in general, have the potential to reduce both fuel consumption and emissions, their medium-term future on the European market seems questionable. The main drawbacks are high system complexity, higher purchase and operating costs, increased vehicle weight, and limited installation space. At the same time, they do not permit completely emission-free operation.

Electric traction drives in automotive engineering exhibit a variety of designs and mounting positions. Regarding the installation position, completely new configurations are possible, such as wheel hub motors. All-wheel-drive systems and torque vectoring are comparatively easy to implement using multiple motors. Functional integration can also be intensified. The spectrum ranges from a partial integration of motor and transmission to fully integrated systems including electric motor, gearbox, and power electronics. The functional unit can even be supplemented with axle components to form a ready-to-install e-axis. This variance in available systems is visualized in Fig. 8.4.

Just as with internal combustion engines, different types of electric motors are relevant for automotive applications. Current commercial use focuses on induction motors, permanently excited synchronous motors, and externally excited synchronous motors. Induction motors have a simple design and are easy-to-manufacture. In automotive applications, squirrel-cage rotors are relevant, in which the stator field induces a magnetic field in integrated aluminum or copper bars in the rotor. As a result, the rotor follows the rotating

Fig. 8.4 Overview of various forms of function integration of electric motors (illustration based on Schaeffler Technologies AG (2014, 2023) and ZF Friedrichshafen AG (2017; 2023))

magnetic field in the stator with a delay. Disadvantages of this motor type are lower efficiency and the reduced volumetric and gravimetric power density.

Permanently excited synchronous motors, also called permanent magnet synchronous motors, offer the best efficiency and gravimetric torque density, as well as favorable reliability and packaging characteristics. The main reason for these characteristics is the absence of excitation windings in the rotor and the associated losses. However, the use of rare earth permanent magnets has significant drawbacks in terms of cost, environmental footprint, and supply chain risks. In this context, the establishment of recycling processes for rare earth permanent magnets is an important task for the future. The permanent excited synchronous motor has been used in the BMW i3, for example.

Externally excited synchronous motors are based on a similar operating principle, but they use copper windings at the rotor instead of magnets for excitation. This results in slightly lower efficiency and higher packaging requirements, but the flexible adaptation of the rotor magnetic field allows good operating behavior. According to the current state of the art, the power supply to the rotor is often realized via slip rings, which are subject to wear. As a result, slip rings can have negative effects on lifetime and efficiency. In its current fifth-generation drives, BMW uses an optimized system based on slip rings in which harmful dust contamination is retained by improved sealing. An alternative is offered by inductive transmitters, which are currently gaining interest in the market (Fig. 8.5).

In addition to the types mentioned, the switched reluctance machine and the axial flux motor also show potential for use as traction motors. The switched reluctance machine is currently the subject of increased research activity as an alternative to the permanently excited synchronous motor. It offers high efficiency without the use of rare earth elements, but the control of the motor is more complex. Axial flow machines are also of research interest because they offer high power density in combination with a small packaging impact. Although permanent magnets are used, their quantity is reduced. Apart from these new motor types, there are trends toward higher operating voltages (800–1000 V), higher motor speeds, and optimized cooling concepts. Another challenge is the electrification of the medium- and heavy-duty segments.

As an alternative to diesel, gasoline, or batteries, hydrogen (H_2) has a calorific value of 33 kWh/kg and can be used as an energy storage medium. The hydrogen can then be used by fuel cells or even internal combustion engines without producing CO_2 emissions. While hydrogen combustion is expected to play a greater role in heavy-duty and off-highway applications, fuel cell electric vehicles (FCEVs) could be a complementary technology for

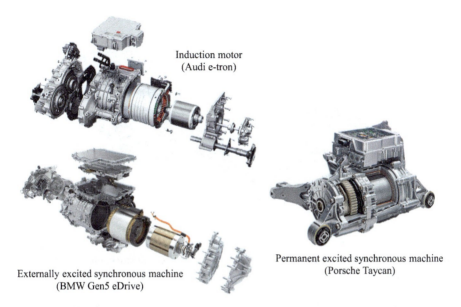

Fig. 8.5 Series drivetrains based on different types of electric motors (illustration based on AUDI AG (2019), Dr. Ing. h.c. F. Porsche AG (2021), and BMW AG (2020))

zero-emission long-distance individual mobility. These vehicles include conventional electric drives in addition to the fuel cell stacks and storage systems. Range is less of an issue, since hydrogen can be fueled in 3–4 min. Hence, hydrogen vehicles combine "the best of both worlds": They offer all the advantages of electric driving, such as instantaneous acceleration and a smooth, silent, and emission-free ride, combined with the convenience of the fast refueling associated with combustion engine vehicles. Together with its partner Toyota, BMW has many years of experience in the development of FCEVs and has recently launched its second generation of fuel cell systems in the BMW iX5 Hydrogen pilot fleet (see Fig. 8.6).

A perceived drawback is the efficiency of the hydrogen energy chain because of the conversion steps involved (see Fig. 8.7). First, hydrogen is generated from electricity by electrolysis, made transportable either in compressed form (e.g., for transport in retrofitted natural gas pipelines), or cooled down until liquefaction, or in form of a liquid organic hydrogen carrier (LOHC), and finally converted back to electricity by means of a fuel cell in the vehicle.

However, in addition to pure efficiency, overall system aspects must be considered. The comparison in Fig. 8.7 only takes into account the use phase. When considering the whole life cycle, starting with the mining of raw materials, the production of components and systems, the assembly of vehicles,

Fig. 8.6 Powertrain of the BMW iX5 Hydrogen with fuel cell stack, electric motor, and two high-pressure hydrogen tanks (BMW AG 2022)

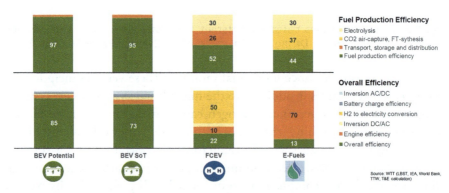

Fig. 8.7 The efficiency of different powertrain options (*BEV* Battery Electric Vehicle, *SoT* State of Technology, *FCEV* Fuel Cell Electric Vehicle) (own illustration based on European Federation for Transport and Environment (2017))

and finally the recycling after the usage, the difference between BEVs and FCEVs is much smaller. In the case of energy generated entirely from renewable sources, conversion losses have no significant effect on the ecological balance.

Also, from an economic and overall energy system point of view, the location and timing of the production of renewable energy must be considered. Self-sufficiency regarding emission-free energy will not be possible in many developed countries, so that they will continue to be dependent on energy imports. Here, hydrogen as a regeneratively produced, chemical energy carrier can make a significant contribution to decarbonization. For example, ideal regions for the production of solar or wind energy are often far away from main industrial areas, requiring hydrogen as an energy carrier for the transport of energy, as electric powerlines have their limitations, especially for long distances. Since the yield of energy production in these regions can be much higher, the conversion losses are mostly compensated by the lower yield of local production of renewable energy in industrial areas (e.g., in Central Europe). In addition, the production of renewable energy depends on the weather conditions, which leads to times of energy surplus when the energy demand is low. In these surplus situations, instead of turning off the production of solar or wind power, the production of hydrogen results in a higher overall system efficiency and also provides additional revenue for the operators of the solar and wind parks.

For the existing fleet of combustion vehicles, synthetic fuels based on renewable electricity can be considered another solution. These so-called e-fuels are produced by reacting hydrogen from electrolysis with carbon dioxide taken from the atmosphere or emission sources. Combustion of e-fuels in conventional ICEs cannot match the superior technical properties of electric drives, and at the same time, they lose a large amount of energy. Therefore, the cost of providing them is extremely high (see Fig. 8.7). While aviation relies on e-fuels to reduce its carbon footprint, these fuels are not currently expected to be relevant for road vehicles in the long term. Exceptions include niche applications, such as racing or vintage cars. For example, Formula 1 will use e-fuels in its hybrid cars beginning in 2026 (Barretto, 2022).

The difficult-to-understand technical characteristics and potentials, the complex factors influencing the performance data, the costs, and, in particular, the environmental compatibility are very difficult to compare objectively among drivetrain choices, even by proven experts (Weigelt, 2022). As a result, despite the clear predominance of electric drivetrains, traditional customers still seem to feel daunted when faced with the necessary change in attitude, and they continue to cling to familiar ICE cars, as reflected in the number of registrations and the fact that governments have to set targets for the number of electric vehicle registrations (Association des Constructeurs Européens d'Automobiles, 2022; Bundesministerium für Umwelt, Naturschutz, nukleare Sicherheit und Verbraucherschutz [BMUV] 2021; Bundesministerium für Wirtschaft und Klimaschutz [BMWK] 2023). Therefore, one of the major challenges is to increase customer acceptance by demonstrating technological reliability.

Mobility must always be considered holistically. In addition to the vehicle and its drivetrain, this includes the infrastructure and the established energy ecosystem, which vary greatly from country to country. These aspects also have a major impact on the overall efficiency and the ecological footprint of the various drive concepts. In the case of electromobility, for example, these are technological disruptions in cell chemistry, the charging infrastructure, and the closing of material cycles for battery raw materials. For hydrogen and synthetic fuels, these include the production and eco-efficient transportation of green hydrogen or synthetic fuels. Unfortunately, this situation challenges incumbent car manufacturers to manage the uncertainties surrounding emerging technologies, while new entrants seize the opportunity to focus on new technologies.

After this brief introduction and comparison of alternative drivetrains, the following two dialogues between experts from research and practice will address the current challenges and opportunities for the future of the drivetrain and the potential of hydrogen in the automotive context.

8.3 Expert Conversation on the Future of Mobility

What is the History of Electric Vehicles at BMW and Where Do You Stand Today?

Franke: It is a real honor and a great pleasure for me to talk with you about the future of mobility. The future of mobility, especially automotive mobility, for me is summed up in the four letters C, A, S, E, which stand for connected, autonomous, shared—I personally would redefine it as sustainable—and, of course, electric driving. This was the mantra when BMW decided, 10 years ago, to design and produce a fully electric car, the BMW i3. I drive an i3 as well.

Zipse: Do you? I hope you enjoy driving it.

Franke: It is still an excellent car, even though it is about 10 years old. It is fully purpose-designed for electric driving, although it started with a small range extender to overcome the range anxiety. This range extender has disappeared, and now BMW has changed its strategy a bit. As far as I can see from the outside, BMW's strategy now is to offer all drivetrain alternatives in every model series. However, fully purpose-designed electric cars would potentially outperform BMW models in terms of better integration, cable harness, compartment design, maybe even efficiency and costs. How will BMW overcome these potential disadvantages?

Zipse: Thank you very much for that really strategic question. What you call CASE, we call ACES, but it means the same thing: autonomous, connected, electric, and shared. In terms of electric cars, which is what we are talking about here, we have not changed our strategy. We have evolved it into the future. We built the first electric car in 2008: an all-electric Mini. It was not for public use, but there was an electric Mini in existence. There was also an electric 2-Series, you know. We experimented with that, and then, we finally made the decision, "Let's build an electric car." At the time, we called it the Megacity Vehicle. It later evolved into the i3. We also said, more or less, this can be an experimental field for car body design, as we had to reduce the weight to increase the range of the car. The i3 became a carbon car because of the lightweight constraints. It is still the only high-volume carbon car in the world to this very day. There is no other manufacturer that has built a car like this.

Franke: What has happened since then?

Zipse: Back then, we were working with the third generation of our battery technology. Now we are in our fifth generation. Back then, we knew—and in the automotive industry you have to look years ahead—that electromobility was coming. Up until then, there was no mandatory use of electric mobility. There was not enough infrastructure, and there was no customer commitment to buy these cars. But because we saw that this was going to happen because of the carbon regulation that was coming, first in the United States and then in Europe, and then also in China, we made a bold move, a very early move. We changed our strategy at that time, not afterward; that is the misconception. We took the change that was coming very seriously. Then, we waited a year and a half, until about 2014, to see how the i3 was doing. We saw competitors coming up and making plans. So, we said, "OK, let's get serious now!" Then, after the i3 experience, we took a very consequent step of electrifying our main architectures. These are not conversion products, but are built on flexible architectures. This took 4 to 5 years, while the outside world had the impression that we had stopped doing electrification, which is not true at all. We prepared for the point in time when electromobility really would take off. That time is now: The iX3 was our first fully electric car after the i3; the iX is the second car. There was a test in Auto Motor Sport the other day, and they tested six fully electric vehicles—purpose-designed vehicles. Other than the iX3, there were vehicles from our German friends, Japanese manufacturers, and so on. They were all there. The iX3, which is built on a flexible architecture, not on a purpose-built architecture, came out on top by a wide margin. First prize! It is the best system overall, and people are buying it.

Franke: So what are the benefits of working with flexible architectures?

Zipse: People do not buy ACES or CASE; they buy cars. People always buy a system, and there is this misconception that you can say, "Well, I built the best battery, and people buy the car because it has the best battery." Even the first i3 with a range extender did not have enough range. So, customers always buy complete cars. There is a misconception that you can only build electric cars on singular, dedicated architectures. That is a mistake. We are now at the point where between now and 2030, electric mobility is going to take off on a large scale, and in a company like ours, you build up entirely new architectures every 10 years, including on the digital side. Now, we have made the decision that, in 2025, the volume for battery-only vehicles in the market will be high enough to develop a dedicated architecture and also to renew not only the electric drivetrain, but also the digital architecture. This is the point where we replace our current architecture with a new architecture. It is not that dedicated electric architectures are better than others. The logic is a different one. And we are ramping up massively. By the end of 2023, we will have launched 12 battery-only vehicles. The next 7 Series will come with four drivetrains. It is the same principle with the iX3, which has won all the tests. We love the next 7 Series. It will be a hit; we are preparing to ramp up quickly, and the volume is there. That was the reason for this decision. It was a transformation of our strategy, not a change of direction.

Franke: Not only the technical tests prove your concept, but also the financial success tells us that your strategy is right. How many cars per year do you need to sell to dedicate a single-purpose-design platform?

Zipse: Usually, we build 6–8 cars on one architecture, for a production volume of about 1 million cars. But, of course, the customer does not see that this is one architecture. You know, the cars built on one platform come in completely different shapes. Nobody would assume that the 4-Series convertible and an X5 are based on the same architecture. But they are.

What Is the Strategy for the Electric Drivetrain?

Franke: Let us take a closer look at the electric drivetrain. As a production engineer, I am impressed by how BMW designs and builds not only combustion engines, but also electric motors. But as a production engineer who focuses on the production of electric motors, I also know that the drivetrain is made up of not only electric motors, but also batteries and power electronics, and of course, it is integrated in a charging network. How will BMW differentiate itself in the future in the other modules of the drivetrain?

Lamp: It is exactly as you say. It is not the single component, but the full drivetrain that makes the difference. That is where we have our experience. As Mr. Zipse mentioned, in the fifth generation, we tried to optimize the whole system between the electric motor, the charging unit, and the battery pack. As you said, the e-motor, in particular, is not only one of the most efficient e-motors, but we have more than doubled the power density of the e-motor between the i3 and what we now see here in the iX. More importantly, in the iX, we are focusing on electrically excited synchronous motors that are free of rare earth metals. This is essential for sustainability. We are very pleased with that. We have gone one step further, by fully integrating the motor, the gearbox, and the power electronics in one housing, which reduces the cost and makes it more compact for integration into different platforms, such as a multi-purpose or a special-purpose platform.

Franke: How do you reconcile this integration with the diversity of your products?

Lamp: The key is that the set of components is built to be scalable across the different models. We can go from about 90 kilowatts up to almost 400 kilowatts with the same concept of the e-motor. The same goes for the other components. The battery is the most significant in terms of cost—80% of the drivetrain costs is related to the battery. Here, we have the same thing: a very simple building block. We design or define the cells in such a way that we can build very standardized modules that are then assembled into different battery packs.

Franke: Even the battery cells?

Lamp: We specify the cells ourselves. So, we have never built—or, let us say, bought—cells off the shelf. Even when we built the i3, which was one of the first large automotive battery cells, Samsung was producing at that time, and we had the opportunity to build up our competence to be on an equal footing with the cell manufacturers. So we worked together to find the best solution for our standardized building block for a battery. We have different module sizes, but the basic way we build it this is the same. We can make different numbers of cells with the same equipment.

Franke: In my professional career in business and industry, I have learned that it is all about core competencies. It is not only the design or the definition of the product, the module, and the parts, but also the production and even the production technology and the tools. The other competitors produce and design their battery cells; they use silicon carbide power electronics, for example. What modules will BMW produce in the future to claim this as a core competence and differentiate itself from its competitors?

Zipse: You should try to build up a core competence if you are likely to have a monopoly or an oligopoly in the market: Then, customers are very dependent on you, as they cannot choose. Or it would be an option if it differentiates you from your competitors, or if your development speed can be significantly faster than in the outside industry. If you can answer yes to these questions, you can build batteries from scratch yourself. But this is not the case. The battery is a rapidly evolving technology.

Franke: Can you elaborate on what these different aspects mean for the battery?

Zipse: First of all, there is no monopoly out there, not even an oligopoly out there. There are a lot of competitors: Europeans, Koreans, Japanese, and Chinese. There is a big worldwide industrial network that is emerging. So, right now, we have four major battery suppliers. We are not dependent on anyone, and we do not have the need to build the batteries ourselves, because with four suppliers, our demand is in good hands. You must have the right contracts, of course. That is a negotiation skill.

The second thing is the speed of development. You increase your development speed when you have more than one supplier. In 2019, we built our own research and development (R&D) center here, where we developed our own competencies. There are more than 2000 parameters in the cell. The right combination determines the performance of the cell. Now comes the interesting part. What happens with a lithium-ion cell? First, you try to optimize the cathode by increasing its performance. You increase the nickel (Ni) content, you increase the manganese content, and you increase all the ingredients of the cathode. But what happens if you increase the energy density on the cathode? You have to change the anode as well. You go away from graphite to silicon–graphite or something like that. Once you have that, the third step is the performance of your electrolyte, and then, the fourth step, which we may do in this decade, is to go from a liquid electrolyte to a solid electrolyte.

Franke: And that rapid pace of development requires flexibility on the supplier side.

Zipse: Exactly. This is a rapid development path from the cathode to the anode and then to electrolyte. It may happen that you need other suppliers along the way. Nevertheless, we are very careful not to invest a lot of time in one technology. In 2021, you might have had the latest state-of-the-art technology and have been well advised to use it, but in 2026, you might need a different technology. There is a big difference between a fluid electrolyte and a solid electrolyte.

Why are we investing? Because there are many competitors out there, and they are extremely competent. We can build up partnerships, as for our business model. It is much better to use the market than to invest in developing specific technologies ourselves. We invest wherever we have an integration task, where all the technologies come together in the car. We refer to this as HEAT, which stands for the German acronym "Hochintegrierte Elektrische AntriebsTechnik" (meaning Highly Integrated Electric Drivetrain), a combination of the electric drivetrain and the clutch end zone.

Franke: I see the arguments with the battery in a similar way to the semiconductor market 30 years ago. In Europe, we thought that we had to build our own semiconductor production capacity. Now, nobody is talking about that. We even have an oligopoly here with a few processor manufacturers worldwide. Taiwan Semiconductor Manufacturing Company (TSMC) makes 80% of the processors. But we are okay with that. So why did you decide to design the electric motor yourself, but not the power electronics? Why don't you get into the production of power electronics, given how important it is for efficiency and dynamics and that it is integrated into the electric motor?

Zipse: Let us take a look at our internal combustion engines. We produce only 5% of the work content of combustion engines ourselves, even though we are called the Bayerische Motoren Werke: 95% is produced by suppliers. We do not make pistons, we do not build turbochargers, nothing. There is a thriving market out there. We are the system integrators, and we believe that we still build the best engines in the world, even with very little value-added in-house content.

The assumption that in-house production gives you a higher competence is simply wrong. Our experience is very different. You build up your own competence when you operate in a monopoly situation. Take press shops, for example. Building those up is very difficult, a super high investment: One costs 100 million euros. There are not many market players out there who build press shops. It is not high-tech, but you have a very clear oligopoly situation—and you are completely dependent on press shops. It is a simple technology, and it makes sense to invest in ourselves.

That is the situation we have. Of course, in our purchasing department, we always look at the market situation. If we cannot increase our speed and we do not have a shortage of supplies out there, we rather tend to buy.

Franke: And when will you bring your first silicon carbide power electronics to market? This can save 20% of the energy, as competitors have already successfully demonstrated.

Zipse: Launches always depend on the introduction of a new architecture. We live in a 7-year cycle, so some OEM will always be earlier than another. We will integrate them into our next architecture. It is a normal process that a competitor may start earlier because their architecture will be new. I am not afraid of that.

What Does the Future of Mobility Look Like?

Franke: I see. Let us move on to the next topic. We are going to start a new course called "Future of Mobility." You call it ACES. The competition is extremely strong in autonomous driving technologies: sophisticated sensors like stereo cameras, radar, lidar, and ultrasonic sensors. In our department, we buy powerful control hardware from Nvidia or Mobileye, using graphical or neural processing units on artificial intelligence (AI)-based software trained on millions of miles driven and giant data infrastructure. The major competitors, like Google, Waymo, and Apple—or even Tesla in the United States or Huawei in China—control several of these key technologies: sensors, AI, data infrastructure, and so on. How will BMW maintain its technological leadership in this area over the long term?

Zipse: I would add connectivity to that list. We want to be the technology leader, not only in electric drivetrains, but also in assisted driving. That will come in stages. But, in addition to technology, there are two important factors that we have to bring into this equation. It is not only the availability of technology, it is not only competence, and it is not only the technological capability. For these, we are already there: We have Level 5 cars driving out there in our test centers. But two other factors are critical. First, what do the legislators do? Do they allow you to sell a car with autonomous driving capabilities? And what are the specific requirements in Germany for them? For example: You can have Level 3, but only up to 60 km/h. Second, you have to ask yourself—and that is the final part of the equation—Is this a business case? Because between Level 2 and Level 3, there is a price tag of more than 10.000 euros, which is a lot of money even for premium customers. You have to check all three boxes: technological competence, legislation, and a valid business model.

Franke: So you are not afraid of Google or Apple?

Zipse: The companies you mentioned are not building cars. They build technology. Inside the automotive sector, they are not making any money. So the question is: When do you, as a car producer, make specific investments without looking at the contribution margin? It is quite simple: When we bring a car to the market, the cost has to be lower than the price needed to make a positive contribution margin. None of these companies think about contribution margins. What will happen, at least for the next 10 years? We will be at the forefront of assisted driving that goes to a Level 2+ with driver supervision in Germany—which is hands-free, by the way.

People think that only Level 3 is hands-free, but this is not true. You, as the driver, are monitored to see if you are still looking forward, and if you are not, the car will ask you to take over control. Level 3 is completely hands-free, but that is highly restricted around the world. To offer Level 3, you need radar, lidar, and optical sensors like a camera—we are absolutely sure of that. You cannot do Level 3 with cameras alone. But a lidar sensor in a car is very expensive.

Franke: Well, the new Apple iPhone 12 has already has a lidar. Very cheap. I know.

Zipse: This is a matter of progress. We are watching very closely what happens here and at what point of time we see a business model. We are fully aware that there is another race going on, but that is actually happening in a different part of our industry. That race is Level 5. That is an entirely different thing, but it will start in the transportation industry, not in the normal passenger car industry. We will see Level 5 trucks very quickly, possibly in China and the United States. However, those will only drive on highways. They will never enter a city, never enter a normal traffic jam, and never have to turn corners; there are no drivers in there; they just go from point A to point B, like a train. You will see that fairly quickly. In the rest of the industry, you will see people movers, driving at low speed, and so on. Robo-taxis may move people through cities, but this again will be a question of a business case—will the investment in the technology work on a large scale?

Franke: So what does this mean for BMW?

Zipse: The question is: Are you going to participate? We do not build vans today, but it is an important market segment. We are looking very closely at the point at which vans could contribute to our business model. But again, this is not so much a question of technological competence, because in this area, everybody needs partners. Nobody can do it alone. Do you have a business model where you can convince customers to pay a specific price for a specific competence in the car? You will see BMWs with Level 3 on the streets as soon as we are allowed to do so.

What Is the Role of Connectivity?

Franke: I have to touch on at least one last question. We talked about electromobility and autonomous driving, but I think connectivity toward the customer and toward infrastructure is also very important. Modern cars are already connected to the Internet through mobile communications technology, telephone, web conferencing, updated maps with up-to-the-minute congestion information, music, video, entertainment services, and Netflix—you name it. Everything is available in the car.

Equally important is the constant connection of electric cars to the smart energy grid. Dynamic inductive power transfer technology, which means recharging the car while driving it and promises infinite range. Connectivity supports autonomous and convoy driving, stabilizes the electricity grid, and offers new services for individual mobility, such as tolls without pay stations, navigation, fast Internet communication, and so on and so forth. This technology is never discussed. We all talk about batteries and hydrogen, but no one talks about inductive power transfer while the car is moving. How can BMW take advantage of this promising technology?

Zipse: The second letter in the acronym ACES stands for Connected. We talk about that. This car behind us (points to an iX model) is one of the first series vehicle with full 5G capability. Let us go back to autonomous driving: Real 5G is necessary to put the intelligence of autonomous driving outside the car, which is ultimately much cheaper. If you put all the AI, all the components and sensors inside the car, the car becomes too expensive. This is what the Chinese are doing: They invest much more in infrastructure and connectivity than Europe or the United States. The first question they ask is: What is the latency time of the car? Because that determines whether you can do autonomous driving with the infrastructure around you. It is your latency time: A few milliseconds are critical. There are different approaches, but it all comes down to connectivity.

Franke: You just mentioned the connectivity of information. I am talking about connectivity of energy. Why do you not charge your car while you are driving down the highway at 200 km/h? Why do you not use an inductive power transfer (IPT) technology? The technology is there. We have test tracks all around the world. Even 10 years ago, I rode in a bus with 60 kilowatts of power at a speed of 60 km/h. We can easily upscale that to 200 kilowatts and 200 km/h. It is a question of decision, perhaps a political decision, but also a decision by a major car manufacturer who could say to the state or the city of Dubai: We will sell you a million electric cars, and we will prepare the infrastructure with IPT, inductive power transfer. Could this be a new business model for BMW?

Zipse: Good question. Of course, this is a chicken-and-egg problem.

Franke: If you look at it as a project business, which is completely new for BMW, then it is not a chicken-and-egg problem. It would be a project. You install the technology in Dubai all over the city, and you only sell BMWs that are compatible with this project.

Zipse: We would not do anything that we could not scale. We are a global player. We have 140 markets. The only way to be efficient in this industry is to scale.

Franke: Well, you can do it in Dubai, you can do it in London, you can do it in Shanghai, and then you scale it up elsewhere.

Zipse: If we saw the scaling, we would do it. But right now, it is not only about scaling; it is something else. On the A5 highway between Darmstadt and Frankfurt, there is a 10 km stretch for truck catenary charging. I drive there sometimes, and we keep a close eye on it, because it is an excellent solution for trucks.

Franke: But not for passenger cars?

Zipse: Even for trucks, it is not being scaled up in Europe or in Germany. That would be something where we would say, we will put it on the R&D side; we will see what happens. But we would not make a solitary, heavy investment just from BMW to scale that up. Because we do not see that. What could happen before that, especially in China, is that there are completely green cities that are built from scratch. They could be based on a fully integrated solution. In those scenarios, we would offer ourselves as a technology partner. But there are not many cities of a certain size that are built from scratch.

Franke: Thank you very much, Mr. Zipse and Mr. Lamp, for sharing your insights with me. It has been a pleasure for me to discuss these new technologies and strategies at BMW. Thank you very much for this discussion.

8.4 Expert Conversation on H_2 as Fuel of the Future

What Is the Importance of the Hydrogen Strategy?

Wasserscheid: I am really excited to be here and to discuss with you the topic of hydrogen as a future fuel. It seems to me that hydrogen is a very dynamic field at the moment. A lot of scientists, companies, and even politicians have recognized that this fully defossilized energy system that we want to have in 2045 can only work if we have storable energy carriers. The time has come: People are developing hydrogen strategies everywhere. There is a Bavarian hydrogen strategy, there is a German hydrogen strategy, there is a European hydrogen strategy, and my first question is: What about a BMW hydrogen strategy?

Zipse: I have been with the company for 30 years, and now, we only have less than three decades until 2050 to become carbon-neutral. That is not very long. It sounds very long, but it is not very long. By then, at the latest, we need to be at least carbon-neutral, better yet, zero emissions. We still think that one of the main paths toward this is—as we discussed before—electromobility. That is the main route. The charging infrastructure is being built. There is a strong consumer demand for electric cars. It will become mainstream—very quickly, we see that coming. The only question is: What if, at some point, you have to be completely emission-free? Electromobility may only be the right answer in certain circumstances. If you do not have access to charging infrastructure, if you do not have access to renewable energy, you need some form of energy storage.

Wasserscheid: So do you agree that energy storage is a relevant topic?

Zipse: We believe that, for automotive market segments, hydrogen is the best answer. I am not just talking about passenger cars, but especially about buses and trucks, marine, aviation, and so on. So, the applications for hydrogen will be much broader than what we see in electric mobility today. With a view to 2050, we see that for BMW—and we can only speak for BMW with a global market share of 3.4%—hydrogen will be an essential ingredient in that mix of propulsion systems, especially because, already today, unlike Germany, we have quite progressive hydrogen-focused countries, such as South Korea (Hyundai) or Japan (Toyota), which have been pushing this technology for almost a decade in serial production. BMW is a global player, and we see this as an important part of our premium brand strategy.

Wasserscheid: In Germany, I am often asked this question: Where will the green hydrogen come from? People say we need 2.5 times more electricity to produce it, but I see it a little differently. My point is that, today, Germany is an energy-importer—and all analyses also show that this will also be the case for future energy systems. Today, 80% of our energy comes from other countries, and it will be similar in the future. The way we will import energy is mainly through hydrogen: hydrogen derivatives, chemically bound hydrogen, ammonia, LOHC, and whatever you want. The point is: If you have hydrogen as a transport vector to Germany, the efficiency discussion is completely different, because it makes no sense to convert the hydrogen into electricity to charge the battery in your car. It is better to use this hydrogen immediately and directly in mobility, of course first in sectors where batteries have problems (e.g., trucks).

Would you say that we have enough electricity in Germany? Because this is not about Germany, is it? Climate change is global. There are places where it is much easier and more economical to generate renewable energy than here in Germany.

Zipse: Right. We have a global perspective on hydrogen. This is important because only about 8% of our worldwide revenue is generated in Germany. So, we have to look beyond the German hydrogen discourse. And if you take a global perspective, you see many use cases where access to electromobility is lacking. From our point of view, the only emission-free possibility for private passenger cars, apart from battery electric vehicles (BEVs), is fuel cell electric vehicles (FCEVs) that run on hydrogen. There will be plenty of cases where people will not have access to charging infrastructure/electricity. So, hydrogen is a perfect complement to our overall strategy. It obviously does not mean that we are undecided. On the contrary, we are determined because we are not in a shrinking scenario. We do not see hydrogen as a shrinking scenario. Because we produce almost 3 million cars per year, we can afford to have three or four different drivetrains, especially because our whole mindset is about architectures and not around platforms. That is the perfect strategy for us. That is the way forward.

What Will a Hydrogen Car Look Like and What Is Its Advantage?

Wasserscheid: It is a disruptive technological change to move from fossil to renewable technologies. If you are going to build your first BMW hydrogen car, what kind of car and what kind of customer are you targeting?

Guldner: We just have completed the development of our second generation of fuel cell technology and integrated it into the X5, one of our bestselling models. A small pilot fleet of BMW iX5 hydrogen vehicles is currently used worldwide for testing and demonstration purposes—very successfully. Then, we will move on to the next development steps, and customer cars will be ready when the markets are ready for them. Different countries move at a different speeds, and we will see when the right point of time comes, probably before the end of this decade.

Wasserscheid: I drive a hydrogen car myself, as a private car, a Hyundai NEXO, and I am really waiting for the BMW. Do not forget me! I am your first customer!

Guldner: We will call you, and you can come for a test-drive soon (laughs).

Wasserscheid: Perfect! And the experience is that this is a nice way of driving, especially for long distances, because even though the network of fueling stations is still quite thin, it is basically enough if you drive long distances. You pass a filling station every 50 km. That is more than enough.

Let us go into the future, into the year 2045, when Germany is supposed to have zero emissions, according to the new climate laws.

What will be the ratio between battery electric and hydrogen electric then? Because my feeling is that batteries are moving fast today because you already have the scale and the mass production effect. In the hydrogen business, the situation is still different. To give an example, if you want to get a cheap fuel cell today, then the best thing is to buy a NEXO or a Mirai, take the fuel cell out, and throw the rest away. The reason is that fuel cell production has yet to be scaled up. What is your outlook for 2045? Let us just assume that, by then, technology development is in the mass market for all technologies, and it is really an established market for both technologies.

Zipse: The question is, which parameter, which vector do you believe in? Do you think that the people are afraid of too long charging times in electromobility—even in the very best case, it will be longer than at a gas station. People might not want to stand in the dark and wait 10 min. They want to refuel quickly and then go. We do not know that. I think the main driver for hydrogen will come from settings in which you have to drive emission-free and you do not have a charging station. That is the main driver. It is not range or anything like that. It is not even the cost. For example, if we say that we want €5 per kilogram of hydrogen, and that would be the point at which it becomes competitive. It is also when countries, through legislation, no longer allow greenhouse gas emissions from cars—and there will be many in 2040. Not all, but a lot. What happens if you cannot charge all these electric cars?—The issue is not the availability of battery electric cars, but the availability of charging stations.

Wasserscheid: What kind of example do you have in mind?

Zipse: What will you do in rural areas? In densely populated areas, like most of Germany, you can provide enough charging interfaces. When you build new houses, you build in charging facilities. You can do that in cities or in the countryside. But in a very scenic, unspoiled environment, where nature is dominant, you cannot tear up all the roads and build a massive amount of charging infrastructure. You will get a massive political problem. What do you do there? But remember, by 2040, we want to drive emission-free. So, either you stop driving cars to those places, which is one option. Or you have a hydrogen car.

Wasserscheid: But the same goes for the city of Munich, right? If you spend half an hour every evening looking for a parking space, it will certainly be more difficult in the future to find a parking space with a charging station. That could be a similar scenario.

Zipse: A similar scenario, yes. I think that is the most dominant parameter: You will not have enough access to charging points. I am not talking about the problem that the build-up of overall charging infrastructure is not moving fast enough. That is also a challenge. But even in societies where the charging infrastructure is developing rapidly, there are simply places where it is too expensive to install charging systems—places you simply cannot or will not access. It would be like trying to bring public charging to the last house in the Black Forest. You are not going to do it.

Guldner: There is also an economic parameter: There are studies that show that a dual infrastructure—both hydrogen fueling stations and electric charging—is cheaper than just putting everything into electricity, because building the electric charging infrastructure has a nonlinear cost vector, because the more you put out there, the more charging stations, the more expensive it gets. But the hydrogen fueling stations always cost the same. From a practical standpoint, the hydrogen fueling stations already exist—they can be easily integrated into existing gas stations. You do not have to discuss who owns the land, who operates it, and so on. All those things are already in place. It makes it a lot easier to get it out there.

Zipse: That is a good point. It is easier to roll out hydrogen gas stations nationwide.

Guldner: And the synergy is with the commercial vehicles, especially trucks, where we have to roll out a hydrogen fueling station network anyway. It is the same hydrogen.

What Is the Technology Strategy Behind Hydrogen?

Wasserscheid: Let me dive a little deeper into the technology strategy. Hydrogen is a brand-new technology; it is disruptive. What is the value chain? And who will be involved with what kind of service and with what kind of product? I guess that this is also a strategic decision for BMW. When and where do we form alliances with other OEMs for fuel cells—I know about your collaboration with Toyota—or do we source in from classical suppliers like

Bosch, Schaeffler, and so on, who fortunately are also quite active in the hydrogen business? I think it is very encouraging that German companies are recognizing that this kind of business fits very well with what they know, maybe different from making batteries. What is the strategic decision for BMW to have a Unique selling proposition in the hydrogen race, but also to have its own product portfolio within this powertrain of the future?

Zipse: We already touched on this point before with the batteries. I think that in-house competence in all areas is overrated. System integration competence is the most important thing you need to have. Then, you have to have a very strong ability to cooperate. If you have strong partners out there—and we do have partners who have been cooperating with us for many years—then you do not need to have all competencies in-house unless you have an extremely unique technology, which is not the case here. I am sure that with all these hydrogen drivers, there will be a thriving supplier market. There will not be just one fuel cell supplier. The important thing to remember is that the integration possibilities—because we already have an electric drivetrain—are quite simple. At the end of the day, a hydrogen car is more or less an electric car. Our car architectures are designed so that BEVs and FCEVs have a high degree of synergy. This is crucial for our electric strategy.

We have a long-standing relationship with Toyota that we are very happy with. Once we decide on a series model, we have to look at who is the best supplier or the best partner. Regarding a make-or-buy decision, it is not always true that the in-house made decision is the optimal choice. It can be right, but most of the time it is not, because you lose a lot of flexibility.

Wasserscheid: This discussion goes even deeper if you think about the whole hydrogen value chain, with green hydrogen production, hydrogen logistics, fuel cells as one way of hydrogen utilization, and the aspect of industrialization and scaling. People sometimes ask me, "Oh, hydrogen has been around for 100 years! Why is it not there yet?" The main problem is this relatively complex value chain: You have to find suppliers and partners—and that is not easy at the beginning. For example, there are different technologies for on-board storage: compressed, cryo-compressed, liquid, and all of that. It is important that the first movers—Toyota and Hyundai are excellent examples—are able and willing to cover parts of that value chain themselves. Otherwise, the technology will not begin to move.

Zipse: Do you see that change happening yet?

Wasserscheid: I am convinced that we will now build this value chain. We see this in the Bavarian Hydrogen Centre. There are companies that do electrolysis, like Siemens Energy; companies that do different kinds of hydrogen logistics, like Hydrogenious LOHC Technologies; and companies that do fuel cells or fuel cell components, like Bosch and Schaeffler. When all these elements come together, the chain will work. What about your willingness as BMW to contribute your part to building these value chains? Is it just to say, "I have to buy this, this, and that, and then I will prepare a car"? Or is it, "I believe in this technology, so I want to be this kind of enabler like Hyundai or Toyota"? We've seen the Toyota cars at the Olympics—they really have a mission on hydrogen. Is the same true for BMW?

Guldner: Of course. We have already mentioned the value chain, and we work closely with Toyota. We also work with a number of suppliers on the components, and then we do the system integration. Look at the fuel cell system itself: We do the integration ourselves because that is our core competence. Also, with the tank system, the storage system, we buy the tank vessels and so on, but we do the system integration. We are working hard to integrate that into our electric vehicle architectures in the future. That is where our particular competence comes in: Taking the components, putting them together into systems, and then having a powertrain that is really BMW-like, that has the BMW driving dynamics—as you might experience when you come back and visit us again for that test-drive (laughs).

Wasserscheid: I would love to do that. And I understand that you need partners to do that kind of system integration. Where else do you need partners?

Guldner: We rely on other players to build the infrastructure. H_2 mobility has done a great job in Germany, building up the first 100 fueling stations. Of course, we need a Europe-wide network, and we see that coming with the European initiatives. Hopefully, in the next 5 to 10 years, there will be gas stations that will sell green hydrogen at a reasonable price for both passenger cars and commercial vehicles because, at the end of the day, it is the same molecule, unlike diesel, where diesel for trucks is different from diesel for passenger cars. This is very exciting to see, and I am sure that there is no need for us to invest in a hydrogen fueling infrastructure, because it is already happening.

What Is the Impact of Hydrogen Technology on Climate Goals?

Wasserscheid: When we talk about defossilization, green products, and a low carbon footprint, we also have to consider the production process for cars. In Spartanburg, I think you were the first company to show that hydrogen mobility can be very helpful in reducing the carbon footprint of your production by using hydrogen-powered forklifts. To what level can you extend that experience? And to what level do you think that competitors and other companies that move goods from Point A to Point B would adopt this kind of technology?

Zipse: Building a highly sophisticated, innovative, and high-performance vehicle is something no company can do alone. That is why we have strong cooperation partners. We are in a constant synergy with regulators, with policymakers, with other industry players, with suppliers, and with external engineering companies. It is always a concerted effort. It is essential that we make it clear that this is part of our strategy for the future and that we are not going in and out with different suppliers every year. Now is the time when our environment is making a big effort, and the recent Important Project of Common European Interest (IPCEI) is just one example of many. We have never had a billion IPCEIs. This is the first time that Europe is doing this. The United States, China, and Korea are doing the same thing at the same time. So, this is a unique opportunity. I am not saying that this is something that will be rolled out quickly across all segments of the car industry, like electromobility. It will find its place in the upper premium segment, where the first buyers are. We have been working on this for more than 20 years. We know that this is a difficult topic. When you only have two choices—700 bars or -250°C—it is a different question (laughter).

Wasserscheid: This is why people are trying to come up with other technical solutions.

Zipse: We know that. But we still think that in our 25 years of R&D with hydrogen, there has never been a better opportunity than now.

Wasserscheid: Absolutely. That is why I said in my initial statement that these are dynamic times: Politicians are convinced—and I think they are right—that this is a significant path forward. It is a strong element of a future defossilized energy system.

Zipse: BMW is strongly committed to hydrogen. Of course, there is the argument that "you are not focused enough," that "you should only focus on one technology to be productive and efficient." We don't think that is the right way. The synergy is not in the drivetrains; it is in the architectures. The choice of the drivetrain is driven by customer needs. The biggest risk of focusing on just one or two technologies is that you end up in a shrinking scenario because customers behave differently around the world. And we strongly believe that if you want to grow in this industry, you have to do that through market mechanisms and technology openness. If there are not enough customers for a certain technology after a period of time, you stop.

But, to give some examples, we see that diesel is not dead—far from it. In Europe alone, one in five customers still buys a diesel car. It is a thriving market. The petrol engine is still there, also the hybrid: We are the world's largest producer of hybrids among all OEMs. This is a flourishing segment for us, despite all the public discussion. Pure electric cars have also grown steadily in recent years. So all four drivetrains are very profitable. The question is: Will hydrogen be the fifth drivetrain, or will it be another technology? It does not matter—as long as you have thought about the possibility and put it into architectures. That is what we have done and what we are doing. So far, that strategy has been absolutely right, because in terms of the overarching strategy, we know that if a company of this size does not grow, it will run into problems. So we have to grow, and the lever is technology openness.

What Are the Limiting Resources of the Hydrogen Business?

Wasserscheid: At the end the day, the customer decides. I think what many people forget is that a car that goes 700 km and takes at least 20 min to charge is a different product than a car that takes just a few minutes to refuel. There will be customers who appreciate that difference and who are willing to pay a premium for it. But let us assume for a moment that the hydrogen business grows, especially in the heavy-duty sector. What do you think about resource constraints? Do you see—also in comparison with other technologies—resource-related limitations?

Guldner: Well. There are two things to this. One is—as we discussed earlier—the efficiency question. We want to look at the entire value chain from the point of view of resources or raw materials: from the manufacture of the vehicle and the production of all the components and assembly, to the use phase, and finally to recycling. The main material used in fuel cells is platinum, and platinum already has a very high recycling rate. In the next few years, there will be a lot of catalysts coming back from all the other vehicles. So the recycling business for platinum is already working very well. That helps in the long run in terms of reducing the amount of raw materials. The battery is a different story. We are putting a lot of effort into making the way we use our raw materials more sustainable. But the battery recycling business requires more effort, and it is just starting to ramp up.

Zipse: All resources are limited. Always. And the car industry is one of the most resource-intensive industries. The two questions are: First, when resources become scarce, do they become more expensive? And second, when does scarcity become an economic argument for survival? That is why we have defined the principle of "secondary material first" as one of the main pillars, at least for the next 10 years. Because humanity is extracting about 100 trillion tons of ores and resources from the planet. Then you ask people, "How long will that continue?" One thing is for sure. Most resources are finite. You can still use them, but they will become more expensive. Palladium, rhodium,… even steel are getting more expensive. So, it makes economic sense to put secondary materials first, at least as a cornerstone of your strategy. I think that is a wise decision.

Wasserscheid: Yes, it helps that platinum is so valuable and people have already developed processes to refine and recycle it. Plus, the quantities that are used in fuel cells and hydrogen release units are small.

That brings me to my last question: There is one resource that may also be limited. And that is brains. Right? (All laugh and nod.) For all that we have discussed, you need people. People with a different mindset. If you are trained as a person with gasoline in your blood, you are quite likely to fail in this electric world. Or maybe not? So the question is: How do you manage the talent pipeline or attract experts in technologies like batteries, hydrogen, and all these electrified mobility technologies? And how does that change the way a company like BMW is going to operate in the future?

Zipse: You hit the nail on the head, because of course brains are a limited resource. But much more important than the limited brain is the mindset. We find a strong source of power in trying to get the right mindset. We did not find such a big difference between the different drivetrains. An electric car is not that different from an internal combustion car. All of the electrics—apart from the battery—are very "mechanical." We move people who work in the internal combustion engine plant to the Dingolfing plant. It is almost the same skill set. And it is not that different. What stays the same is the continuous progress in every technology. Take digitalization, which in principle is not new to us, but every year something new comes along. Putting the right digital solutions into a car and making a profitable business model out of it, that is the most important question. At the end of the day, your product has to be unique, profitable, and attractive. That is not just a question of qualification or whether we have enough digitalization. It is a question of mindset. This is what we are trying to get the whole company to do, that everyone is part of a larger system that pays into a business model. If you have that mindset, you will get enough brains that want to contribute.

Wasserscheid: I think one important way to achieve this is close cooperation and interaction between universities and BMW, as in our very exciting discussion today. Many thanks for that!

8.5 Beyond Technical Functionality: The Energy Ecosystem around Eco-Efficient Drivetrain Solutions

As shown in the previous sections, different technologies compete in the area of individual mobility for different target groups and markets. While some technologies compete directly with each other, others complement each other for different applications and sectors. In this context, technology openness is a key to achieving climate neutrality. In Europe, e-mobility is expected to dominate the mass consumer market in the future. In other areas, such as heavy-duty transportation, it is not yet clear which drive solution will prevail in the long term. All solutions require massive research efforts in various fields to advance the respective technological maturity level.

The associated technological change requires close cooperation between industry (OEMs and suppliers), research, and politics (international and national). At the same time, consumer acceptance must be promoted. The application potential and user acceptance of alternative drivetrains depend to a large extent on the infrastructure and the energy ecosystem. Future practice and research avenues must therefore focus on these two aspects.

In the field of energy ecosystems, three potential future paths for eco-efficient mobile and stationary energy storage are discussed below. Due to the close interaction of new drive technologies with the associated infrastructure, disruptive energy distribution systems in the automotive context are also presented.

8.5.1 Eco-Efficient Storage of Electric Energy on Board an Automobile

The present status and success of battery electric vehicles is closely linked to a specific battery technology—the Li-ion technology. The reason is the outstanding energy and power density of the Li-ion technology compared with other material combinations (see Fig. 8.8).

Several inventions have been necessary to make Li-ion cells work. In 2019, the pioneering work of John B. Goodenough, M. Stanley Whittingham, and Akira Yoshino conducted in the second half of the last century has been

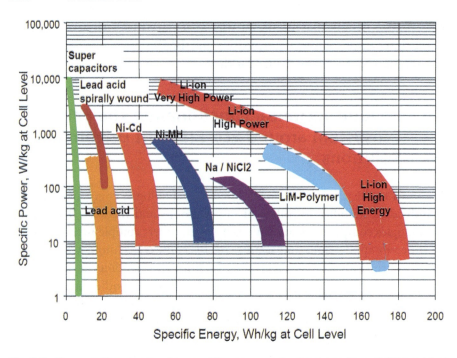

Fig. 8.8 Ragone diagram showing specific energy and power densities of different battery technologies (Note: Status as of 2013 on cell level—today Li-ion expand further to the right; "Ragone plot of various battery technologies with specification at cell level for automotive applications without lithium–sulphur and metal–air batteries." originally published in Budde-Meiwes et al. (2013). Proceedings of the Institution of Mechanical Engineers, Part D: Journal of Automobile Engineering, 227(5), 761–776. https://doi.org/10.1177/0954407013485567; all rights reserved.) (Budde-Meiwes et al., 2013)

honored by the Nobel prize. And it was the need for higher battery capacity for the new consumer electronic devices, like the camcorder, which motivated Sony in the late 1980s to industrialize the Li-ion technology.

The working principle and the subsequent steps to build a Li-ion cell from the incoming materials (powders, solvents, conducting foils, separator, electrolyte, and cell housing) to the final cell is shown in Fig. 8.9. The key materials are the active anode and cathode materials being able to store Li-ions in its inner structure. While graphite is the predominant material for the anode for all applications, the choice of the cathode material (typically a lithium metal oxide) strongly depends on the application and the requirements.

For example, the key performance indicators (KPIs) for automotive applications differ substantially from those of the consumer electronics leading to different solutions. While the lithium–cobalt oxide (LCO) cathode material is used in consumer electronics, LCO is not suitable for automotive

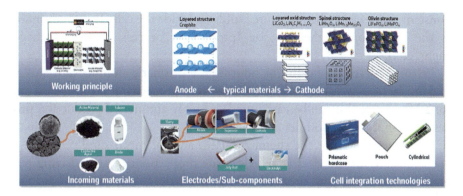

Fig. 8.9 Overview of the working principle of a Li-ion battery cell, the typical set of materials and the typical realization from materials to electrodes and finally to cells (pouch, cylindrical, or prismatic hard case)

applications due to cost and safety reasons. The predominant cathode material in battery cells for electric vehicles today is the lithium–nickel–manganese–cobalt oxide (NMC) chemistry (a layered oxide structure).

Ten years ago, the material composition was the so-called NMC111 (i.e., equal amounts of nickel, cobalt, and manganese [33% of each]). In the future development, the nickel content was continuously increased to 80% in today's so-called NMC811 material (80% Ni, 10% manganese, and 10% cobalt). The benefit is a substantial increase of the specific capacity of the material from about 150 mAh/g (NMC111) to about 210 mAh/g (NMC811). The drawback is that those high-performance materials are thermally less stable, leading to challenges for lifetime and/or safety.

As shown in Fig. 8.9, the active layers of a cell (multiple electrode and separator layers arranged in a stack or jelly roll) can be integrated in different mechanical cell housings. Those are pouch cells, cylindrical cells, and prismatic hard case cells. Each of those has individual weaknesses and strengths, and those are all used in automotive applications.

Pouch cells: The housing consists of a thin composite foil (polymer with an alumina layer in between). The sealing is done by a lamination process. The advantage is its lightweight and high degree of freedom in realizing different form factors (only cell thickness is limited). The disadvantage is the possible diffusion through the laminate seal and the low mechanical robustness.

Cylindrical cell: The housing is mainly steel but could also be alumina. The integration of a cylindrically wound jelly roll gives the best volumetric energy density. Sealing is done by crimping (risk of diffusion but lower than for a pouch) or laser seal. The advantages are the mechanical robustness and constant shape even when pressure builds up inside the cell. The disadvantage is the limitation of the cell size (production process, thermal management, etc.).

Prismatic hard case cell: The housing is a prismatic alumina can. Sealing is done by laser welding. The advantages are the mechanical robustness and long lifetime, and it can be easily used in highly automated production process to manufacture modules and packs.

The battery cell (i.e., the chemistry used) and the mechanical cell concept is responsible for the electric vehicle's core properties of range, driving performance, and charging time. BMW has focused on the prismatic hard case cell as the building block for its battery architecture. The reasons are the mechanical rigidness, the longevity, and the suitability for high-volume and high-quality production of modules and battery packs. BMW has optimized this type of battery over five generations, and it now successfully powers our present fleet of battery electric vehicles (see Fig. 8.10).

In the sixth generation of BMW eDrive technology utilized in the NEUE KLASSE, significant advancements have been made in the cell format and chemistry. The introduction of the new BMW round cell, purpose-built for the electric architecture of NEUE KLASSE models, allows for a remarkable increase in the range of the highest range model, up to 30% (according to WLTP) (BMW Group, 2022; see Fig. 8.11).

Deviating from the fifth-generation prismatic cells, the sixth-generation BMW round cells differ in an increased nickel content on the cathode side. This allows the cobalt content to be reduced. In addition, a notable increase in the silicon content on the anode side contributes to a significant increase in the volumetric energy density of the cell of more than 20% (BMW Group, 2022).

Fig. 8.10 BMW's Gen5 battery cell, module, and pack architecture

Fig. 8.11 BMW's Gen6 battery cell and pack architecture and the resulting improvements of some relevant KPIs (the reference is the present Gen5 battery pack architecture)

In the NEUE KLASSE, the battery system is critical because, depending on the model, it offers flexible integration into the installation space to save space through a "pack to open body" approach that eliminates the cell module level (BMW Group, 2022; see Fig. 8.11).

Moreover, the NEUE KLASSE's battery, drivetrain, and charging technology will operate at a higher voltage of 800 V. This enhancement optimizes energy supply from direct current high-power charging stations, enabling a much higher charging capacity of up to 500 A. As a result, it will take up to 30% less time to charge the battery from 10% to 80% (BMW Group, 2022; see Fig. 8.11).

The BMW Group places strong emphasis on reducing the carbon footprint and resource consumption throughout the production process, starting from the supply chain. To achieve this, cell manufacturers will use lithium, nickel, and cobalt, incorporating proportions of secondary material, that is, raw materials that already exist in the material loop and are not newly mined. In addition, the BMW Group is committed to using only green electricity from renewable sources for battery cell production. Both of these advances are expected to reduce the carbon footprint associated with producing battery cells by up to 60% compared with the current generation (BMW Group, 2022).

Emphasizing the importance of a circular economy in e-mobility, the BMW Group aims to reuse raw materials. Circular loops significantly diminish the demand for new raw materials, reduce the risk of environmental and social standard violations in the supply chain, and lead to substantially lower CO_2 emissions. The BMW Group's active involvement in all stages of a circular battery economy (see Fig. 8.12) underscores their commitment to this approach. Ultimately, the long-term objective is to adopt fully recyclable battery cells (BMW Group, 2022).

For the new generation of BMW battery cells, the raw materials cobalt and lithium will be sourced from certified mines, ensuring transparency over extraction methods and promoting responsible mining practices. The

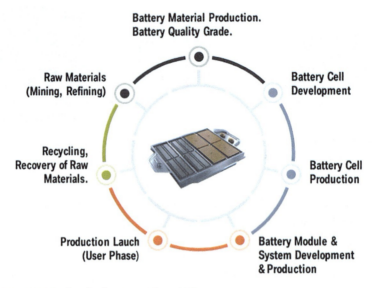

Fig. 8.12 BMW's circular battery value chain

sourcing is carried out either directly through the BMW Group or via the battery cell manufacturer (BMW Group, 2022).

For numerous years, the BMW Group has actively participated in initiatives aimed at establishing standards for responsible raw material extraction and advocating compliance with environmental and social norms through mine certification. This approach not only exemplifies the company's commitment to sustainable business practices but also reduces its dependence on certain resources and suppliers from a technological, geographical, and geopolitical perspective.

For an OEM, the continuous further development of battery systems is mandatory. In the past, the main driver for battery cell development was to increase energy density and hence range of the battery electric vehicle. With the NEUE KLASSE, driving ranges of up to 900 km, depending on the specific vehicle, will be reached. In principle, ranges above 1000 km are technically possible but are, in most cases, neither economical nor ecologically reasonable. It is more important to develop and deliver a product optimized along all relevant KPIs and the needs of the customer. Hence, future battery cell development will diversify and be directed to different areas of the car portfolio (see Fig. 8.13). This will be: a) still the optimization of energy density ("range-optimized"), but equally important, b) the best fit between energy density and cost ("cost-optimized"), and c) the low cost sector for entry models ("low-cost").

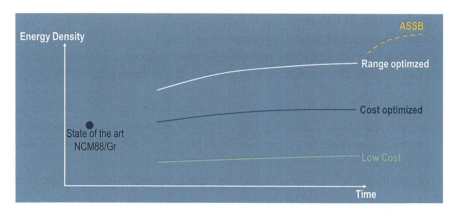

Fig. 8.13 Future diversified development directions for battery cell technology

Since 2008, the BMW Group has been progressively cultivating its expertise in battery cell technology. This expertise has been consolidated at the BMW Group's Battery Cell Competence Centre (BCCC) in Munich since 2019. Encompassing the entire value chain, from R&D to battery cell design and manufacturability, the BCCC serves as a hub for translating cutting-edge battery cell innovations into practical applications swiftly and effectively. To this end, the BMW Group collaborates with a diverse network of approximately 300 partners, including established companies, start-ups, and research institutions. The insights acquired through these collaborations undergo validation at the Cell Manufacturing Competence Centre (CMCC) located near Munich in Parsdorf (BMW Group, 2022).

This competence and continuous effort are needed to ensure that the best possible and most eco-efficient battery technology is offered to the customer.

8.5.2 Eco-Efficient Hydrogen Storage and On-Demand Electrification

Energy from renewable sources enables a climate-friendly supply of electricity, heat, and alternative fuels. However, due to the volatile nature of renewable energy sources, technologies are urgently needed to make renewable energy storable, transportable, and globally tradable to link privileged production sites with centers of consumption. One solution is energy storage in the form of hydrogen, which can be produced by electrolysis of water with renewable energy. If the energy stored this way is needed again, the hydrogen can be used to generate electric energy in a fuel cell, with water vapor as the only additional product. In this way, a CO_2-free energy system can be established.

In the future of defossilized and fully emission-free energy systems, hydrogen technologies will play a very important role. They will provide solutions for applications where battery technologies cannot be used for reasons of cost and practicality. This applies, in particular, to the following areas:

1. Applications where hydrogen is not needed as an energy carrier, but as a "green" reduction equivalent (e.g., the steel industry) or as a reactant for chemical reactions (e.g., for the valorization of CO_2, biomass, or polymer waste).
2. Applications where the amounts of energy stored and transported are exceptionally high and where the number of loading cycles per year is small (typically <20). Examples include the seasonal storage of large amounts of electrical power and the global logistics of low-cost, green energy from global high-yield solar and wind locations for use in industrialized regions with high consumption.
3. Applications in the field of zero-emission, heavy-duty, and long-range vehicles (e.g., for the propulsion of sea ships and river barges, trains, trucks, coaches, and commercial vehicles in the agricultural, forestry, and mining sectors), but also for long-range passenger cars.

One challenge, however, is that elemental hydrogen (H_2) has only a very low energy density at ambient conditions. For storage and transport, hydrogen is therefore stored as a gas under high pressures of up to 700 bar or liquefied at temperatures below minus 250 °C (Preuster et al., 2017). Concerning the transportation of hydrogen, leveraging established fossil fuel infrastructure—specifically, natural gas pipelines—offers noticeable advantages. Utilizing existing infrastructure proves to be more cost-effective and resource-efficient (U.S. Department of Energy, 2023). However, while blending hydrogen into natural gas to transport a gas mixture is comparatively feasible for modest of proportions of hydrogen, converting the gas grid to distribute only hydrogen presents more complex technical, legal, and policy-based challenges (Jayanti, 2022). Furthermore, on an international scale, hydrogen transportation approaches that rely on molecular hydrogen demand the construction of new, considerably expensive infrastructure. Given this context, researchers are exploring alternative methods for hydrogen transportation that extend beyond merely repurposing existing pipelines. Current research and development work at FAU is therefore aimed at establishing innovative hydrogen storage and logistics approaches that are highly compatible with the existing infrastructure for the currently utilized fuels. This infrastructure compatibility offers the chance for a much faster introduction of hydrogen-based clean energy technologies on a system-relevant and global scale.

To realize this compatibility with existing energy infrastructures, the elemental gaseous hydrogen is bound to a carrier molecule in a heat-producing hydrogenation reaction. This creates a hydrogen-rich form of the storage system, the loaded storage compound, which can be easily stored and transported in a liquid or liquefied form. On demand, elemental hydrogen can be released from the charged storage compound in a reverse, heat-consuming dehydrogenation reaction. In this process, the discharged storage material is formed again and can be used for another hydrogen storage cycle. Reaction accelerators, the so-called catalysts, play a decisive role in the described storage and release reactions. They accelerate the rates of reaction and ensure that the desired hydrogenation and dehydrogenation reactions take place with the highest possible selectivity.

A technically very promising example of this approach is the so-called liquid organic hydrogen carrier (LOHC) technology, in which molecular hydrogen is reversibly bound to an aromatic liquid compound (Preuster et al., 2017). Research contributions of the FAU in the last decade have shown that the aromatic compound benzyltoluene is particularly suitable as hydrogen-lean storage compound (Jorschick et al., 2017; Rüde et al., 2022). Benzyltoluene is a readily available industrial product and has been applied since the 1960s as heat transfer oil. Many properties of the compound are known and very well suitable for its application in hydrogen storage, such as its high thermal stability and the high intrinsic safety of the compound. Based on this LOHC system, FAU researchers have developed a hydrogen storage and transport technology that has been commercialized in the meantime by the FAU spin-off Hydrogenious LOHC Technologies GmbH (www.hydrogenious.net). Since its foundation in 2013, Hydrogenious has developed into a global technology leader for hydrogen storage using LOHC technologies and today has 200 employees.

Interesting alternatives in the field of chemical hydrogen storage include the reversible chemical binding of hydrogen to the gases CO_2 or N_2, which also leads to liquid (methanol and diesel) (Artz et al., 2018) or liquefiable (ammonia and dimethyl ether) (Schüth et al., 2012) hydrogen-rich storage compounds. These compounds can be split on demand to produce hydrogen or used directly as chemicals or as climate-neutral combustion fuels (see Fig. 8.14).

An important difference between these CO_2/N_2 concepts and the LOHC technology is that LOHC-released hydrogen is of sufficient quality for re-electrification in a fuel cell after condensation of the liquid carrier. By contrast, splitting ammonia, methanol, or dimethyl ether for hydrogen production

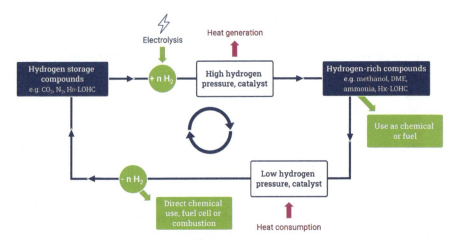

Fig. 8.14 Illustration of the working principle of chemical hydrogen storage

leads to gas mixtures of hydrogen and nitrogen or carbon dioxide, respectively. These mixtures have to be separated to obtain pure hydrogen for fuel cell operation. While the LOHC carrier compounds are transported in the hydrogen-rich state from the energy-rich location to the energy consumer and in the hydrogen-lean state back for recharging, the equivalent storage cycle is typically closed for CO_2- or N_2-based hydrogen storage technologies via the atmosphere. At the energy-rich location, CO_2 or N_2 is extracted from the atmosphere, and the same compounds are released to the atmosphere after hydrogen splitting and separation at the energy consumption site.

Overall, the technologies for chemical energy storage, and for the LOHC technology in particular, offer clear advantages over batteries and physical storage of elemental hydrogen if the stationary energy storage of large amounts of energy, global energy transport, and emission-free heavy-duty mobility are the focus. Chemical hydrogen storage can be realized at ambient temperature and ambient or low pressure to provide very high energy densities. Since the loaded storage material can be easily handled like today's fuel, the existing infrastructure for liquid energy carriers (tankers, tank wagons, and tank farms), which is accepted by the population and has proven itself over many decades, can be further used. There is no need to build expensive new supply infrastructure, nor does the hydrogen require complex cooling or compression. Most interestingly, these chemical hydrogen storage technologies are exportable and can also be used in countries whose gas and electricity distribution infrastructure has so far been poorly developed (Hank et al., 2020).

8.5.3 Eco-Efficient Electricity Distribution via an Electrified Road Infrastructure

The glaring disadvantage of the limited storage capacity and charging capacity of batteries cannot be overcome in the short term. Another major challenge at present is the expansion of the charging infrastructure. According to statistics from the German Federal Network Agency, there were 63,806 normal charging points and 12,755 fast charging points in Germany as of December 01, 2022 (Bundesnetzagentur, 2023). The German government plans to increase this number to 1 million publicly accessible charging points by 2030 (Bundesministerium für Digitales und Verkehr [BMDV], 2022). The European Union plans to install one charging station every 60 km along major traffic routes (European Parliament, 2022). The deployment of bidirectional charging and smart grids will create further synergies.

Other technologies can complement this strategy. For example, instead of storing energy in massive batteries and carrying them in the car, electric energy can be transferred directly from electrified roads to parked and even moving electric cars. Inductive power transfer (IPT) automatically starts the charging process when the vehicle is parked over a charging coil. By installing this technology on long-distance roads, the energy needed to drive can be provided continuously, the concentrated grid load caused by ultra-fast charging is reduced, and the size of the batteries can be significantly reduced. As a result, electric cars can become lighter and less expensive, and even heavy trucks can be driven electrically, efficiently, and with zero emissions for an unlimited driving range without additional recharging. Conversely, by eliminating the need to recharge at rest stops, the investment requirement for the immensely expensive fast-charging columns is decreased, the space required for charging cars is reduced, and travelers no longer waste time recharging their batteries on long-distance trips.

Again, after BMW introduced the i3 as the first purpose-designed electric car in Germany, with the plug-in-hybrid 530e, BMW was the first car company to bring wireless power transfer to the market in series production.

The primary coils (see Fig. 8.15), consisting of concentrically wound copper strands, are installed under the road surface in parking lots or on roads at intervals of about 1 m. A magnetic field pulsed at 85 kHz excites an electric voltage in the secondary coils, which are mounted on the underbody of the vehicles. More than 20 kW of power can be transmitted per coil, which is sufficient even at high speeds and under normal conditions for propulsion and simultaneous battery charging. Higher power requirements, such as for

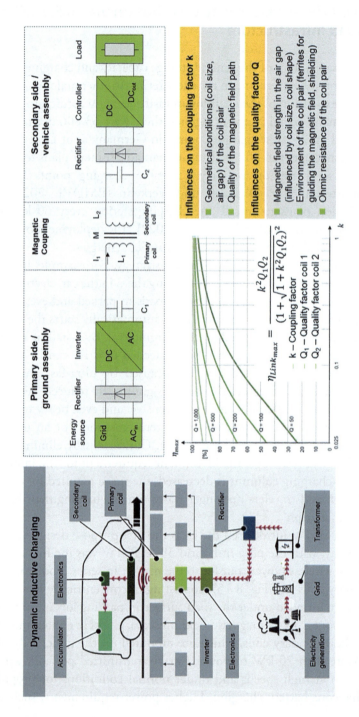

Fig. 8.15 Inductive charging is based on resonant magnetic flux coupling (own illustration based on Loisel et al. (2014))

trucks and buses, can be met by installing multiple secondary coils. With a precisely tuned oscillating circuit and excellent primary and secondary coil qualities, energy transfer efficiencies of over 90% can be achieved, even surpassing the efficiency of previous high-performance conductive energy transfers, since the additional electrochemical energy conversions in and out of the battery can be eliminated.

Since the magnetic fields of the primary coils are basically harmless to living beings and are only activated when a secondary coil system is coupled, the road traffic infrastructure can be electrified without hesitation. However, it will be necessary to educate the public in order to dispel any reservations they may have. The automatic identification of vehicles in the energy system, comparable to the registration of a cell phone in the mobile network, makes the payment process more convenient. The permanent connection of electric vehicles to the smart grid while parking and driving enables the use of vehicle batteries for an effective stabilization of the energy grid (vehicle-to-grid). Even if only half of all German cars were to be converted to electric drives, about one terawatt-hour (TWh) of storage capacity would be available (Loisel et al., 2014). (This is equivalent to about 25 times the capacity of all of Germany's hydroelectric storage power plants.) The introduction of this technology will, of course, require significant investment to retrofit existing infrastructure and expand the electrified road network. The additional infrastructure costs for electrified roads are estimated to be about €1 million per road kilometer (KTH and QiE, 2019).

Compared with other technologies, large-scale inductive charging is still a relatively nascent field. It brings to the fore fascinating research questions pertaining to its potential applications and development. Given the need for technology openness to accommodate various use scenarios, exploring these research avenues can help assess both the technological viability and economic feasibility of this charging method and whether it could complement other charging technologies in the future.

8.6 Conclusion

The mobility sector is currently in a state of uncertainty. Many market players are facing challenges due to technological change, increasing regulation, changing customer behavior, and the emergence of new competitors. Established technologies that have dominated for many years are losing importance and are being replaced by alternative technologies. The most obvious change in the powertrain is the ongoing replacement of the internal combustion engine and the substitution of fossil fuels. This major technological change is necessary to enable a completely emission-free future in the context of climate change, but is also motivated by environmental and health protection. Accordingly, there are strong interactions with the megatrends of decarbonization and sustainability.

The evolution is sequential. Bridging technologies such as hybridization facilitate the transition, while other technologies compete directly with each other as long-term solutions. At the same time, complements are possible for different applications and sectors, such as battery electric mobility and hydrogen as the fuel of the future. It can be assumed that battery electric vehicles will form the backbone of emission-free individual transportation in the future. Regeneratively produced hydrogen, which can be used in fuel cells and internal combustion engines, can provide a complementary solution. Potential applications include long-distance, heavy-duty, or off-highway transportation. In any case, there is a close interaction between the drive technology and the associated energy ecosystem. Renewably generated electricity and hydrogen must be available in sufficient quantities and at attractive economic conditions. In addition, distribution and convenience for the end user must be ensured. This requires a charging and refueling infrastructure that is as capable as the current one for fossil fuels. The European Union's targets of charging points every 60 km and hydrogen refueling stations at least every 200 km along major transport routes and in every city are a step in the right direction. The technology of inductive charging while driving might be another attractive option. In any case, massive investment in infrastructure is essential.

Other automotive megatrends include connectivity, autonomous driving, and mobility as a service (Gall & Sieper, 2021). These developments, which can be summarized under the term ACES, are mostly not directly related to the drive technology used, but benefit from the increasing electrification and digitization of vehicles. As a result, there is the potential to increase safety and comfort while reducing the environmental footprint. At the same time, the technological advances are having a major impact on the OEMs' and suppliers' businesses. Companies that are unable to respond to these developments and uncertainties will face major problems. However, this challenge can also be seen as an opportunity. New markets, products, and business areas are emerging. At the same time, the general public benefits significantly from sustainability, zero emissions, and increased safety.

So how can technologies for future contribute to sustainability?—We would like to highlight five takeaways from this chapter that invite further discussion:

1. The car of the future will be ACES—autonomous, connected, electric, and shared. Each of these characteristics has the potential to promote sustainability if well integrated.
2. As each technology has specific advantages and limitations, the transition to zero-emission mobility will not be based on one technology, but on a mix of technologies that can drive decarbonization across different use cases and contexts. Guided by clear decarbonization targets, technology openness can accelerate learning, experimentation, and flexibility.
3. Hydrogen solutions can be a solution to address the shortcomings of a pure electric vehicle. By reaping its advantages in terms of international energy flows, energy storage, range, and charging infrastructure, it can be an important complement to electromobility.
4. Zero-emission drivetrain technologies require an enabling energy ecosystem that includes not only sufficient renewable energy generation, but also the necessary charging infrastructure. In addition to electric charging stations, innovations in this charging ecosystem can include LOHC-based hydrogen transport and hydrogen delivery at existing gas stations as well as innovative forms of inductive on-road power transfer with the potential to drastically reshape current electric mobility.
5. Implementing technologies for the future requires not only engineering knowledge, but also the right mindset and the ability to partner with other stakeholders. In addition, market-based diffusion of green technologies requires successful business models. From a business model perspective, the system integration of technologies into viable solutions is more important than technologies alone. The implementation of such system integrations benefits from partnerships with others.

Technologies for the future are needed to disrupt the fossil-fuel-based status quo and to develop the products and value chains for zero-emission mobility. This chapter has thus concluded our discussion of the different individual steps on the Road to Net Zero. At the same time, it has shown that technology is related to strategy, products, value chains, and much more. Emphasizing this interplay between the different contributions of this book, the following and final chapter (Chap. 9) *The Road To Net Zero and Beyond* weaves together important threads of our previous discussions and concludes this book with a look into the future of collaborations that can drive the sustainability transformation.

References

AUDI AG (Ed.) (2019). Audi e-tron Bauteile E-Antrieb (Animation). Retrieved July 09, 2023, from https://www.audi-mediacenter.com/de/videos/video/audi-e-tron-bauteile-e-antrieb-animation-4846

Artz, J., Müller, T. E., Thenert, K., Kleinekorte, J., Meys, R., Sternberg, A., Bardow, A., & Leitner, W. (2018). Sustainable conversion of carbon dioxide: An integrated review of catalysis and life cycle assessment. *Chemical Reviews, 118*(2), 434–504. https://doi.org/10.1021/acs.chemrev.7b00435

Association des Constructeurs Européens d'Automobiles. (2022). Electric Vehicles: tax benefits & purchase incentives: In the 27 member states of the European Union (2022). Retrieved June 26, 2023, from https://www.acea.auto/files/Electric-Vehicles-Tax-Benefits-Purchase-Incentives-2022.pdf

Barretto, L. (2022). *Formula 1 on course to deliver 100% sustainable fuels for 2026*. Formula One World Championship Limited. Retrieved June 26, 2023, from https://www.formula1.com/en/latest/article.formula-1-on-course-to-deliver-100-sustainable-fuels-for-2026.1szcnS0ehW3I0HJeelwPam.html

Blois, M. (2022, June 23). Umicore wants to build the world's largest battery recycling plant. Chemical & Engineering News (C&EN). Retrieved June 26, 2023, from https://cen.acs.org/environment/recycling/Umicore-wants-build-worlds-largest/100/i23

BMW AG (Ed.) (2020). BMW Gen5 eDrive. Retrieved July 09, 2023, from https://www.press.bmwgroup.com/deutschland/photo/detail/P90392490/bmw-gen5-edrive

BMW AG (Ed.) (2022). Brennstoffzellsystem. Retrieved July 09, 2023, from https://mediapool.bmwgroup.com/download/edown/pressclub/publicq?dokNo=P90476503&attachment=1&actEvent=image

BMW Group. (2022, September 09). *More performance, CO2-reduced production, significantly lower costs: BMW group to use innovative round BMW battery cells in NEUE KLASSE from 2025* [Press release]. München. Retrieved July 20, 2023, from: https://www.press.bmwgroup.com/global/article/detail/T0403470EN/more-performance-co2-reduced-production-significantly-lower-costs:-bmw-group-to-use-innovative-round-bmw-battery-cells-in-neue-klasse-from-2025?language=en

Budde-Meiwes, H., Drillkens, J., Lunz, B., Muennix, J., Rothgang, S., Kowal, J., & Sauer, D. U. (2013). A review of current automotive battery technology and future prospects. *Proceedings of the Institution of Mechanical Engineers, Part D: Journal of Automobile Engineering, 227*(5), 761–776. https://doi.org/10.1177/0954407013485567

Bundesministerium für Digitales und Verkehr (BMDV). (2022). Masterplan Ladeinfrastruktur II der Bundesregierung. Retrieved June 26, 2023, from https://

bmdv.bund.de/SharedDocs/DE/Anlage/G/masterplan-ladeinfrastruktur-2. pdf?__blob=publicationFile

Bundesministerium für Umwelt, Naturschutz, nukleare Sicherheit und Verbraucherschutz (BMUV). (2021, September 29). Förderung der Elektromobilität durch die Bundesregierung. Retrieved June 26, 2023, from https://www.bmuv.de/themen/luft-laerm-mobilitaet/verkehr/elektromobilitaet/foerderung

Bundesministerium für Wirtschaft und Klimaschutz (BMWK). (2023). *Electric mobility in Germany*. Retrieved June 26, 2023, from https://www.bmwk.de/Redaktion/EN/Dossier/electric-mobility.html

Bundesnetzagentur. (2023, February 2). Elektromobilität: Öffentliche Ladeinfrastruktur. Retrieved June 26, 2023, from https://www.bundesnetzagentur.de/DE/Fachthemen/ElektrizitaetundGas/E-Mobilitaet/start.html

Doppelbauer, M. (2020). *Grundlagen der Elektromobilität*. Springer Vieweg Wiesbaden. https://doi.org/10.1007/978-3-658-29730-5

Dr. Ing. h.c.F. Porsche AG (Ed.) (2021). Hochspannung - Porsche-Elektromotoren erklärt. Retrieved July 09, 2023, from https://newsroom.porsche.com/de/2021/innovation/porsche-elektromotoren%2D%2Dtaycan-hochspannung-christophorus-398-23885.html

European Environment Agency. (2019, December 11). Cumulative global fleet of battery electric vehicles (BEV) and plug-in hybrid electric vehicles (PHEV) in different parts of the world. Retrieved June 26, 2023, from https://www.eea.europa.eu/data-and-maps/figures/cumulative-global-fleet-of-battery

European Federation for Transport and Environment (Ed.). (2017, July 25). *Renewable electricity is a must to decarbonise land freight transport*. https://www.transportenvironment.org/discover/renewable-electricity-must-decarbonise-land-freight-transport/

European Parliament. (2022). Europaabgeordnete fordern Ladestationen für E-autos alle 60 km. Retrieved June 26, 2023, from https://www.europarl.europa.eu/news/de/press-room/20221014IPR43206/europaabgeordnete-fordern-ladestationen-fur-e-autos-alle-60-km

Gall, C., & Sieper, J. (2021, August 4). Wie Unternehmen sich auf die Mobilität der Zukunft vorbereiten können. Retrieved June 26, 2023, from https://www.ey.com/de_de/automotive-transportation/zukunftstrends-der-automobilindustrie-im-ueberblick

Hank, C., Sternberg, A., Köppel, N., Holst, M., Smolinka, T., Schaadt, A., Hebling, C., & Henning, H.-M. (2020). Energy efficiency and economic assessment of imported energy carriers based on renewable electricity. Sustainable Energy & Fuels, 4(5), 2256–2273. https://doi.org/https://doi.org/10.1039/D0SE00067A

International Energy Agency. (2022). *Global energy-related CO2 emissions by sector*. Retrieved June 26, 2023, from https://www.iea.org/data-and-statistics/charts/global-energy-related-co2-emissions-by-sector

International Energy Agency. (2023). *Transport: Improving the sustainability of passenger and freight transport*. Retrieved June 26, 2023, from https://www.iea.org/topics/transport

Jayanti, S. E.-P. (2022). Repurposing pipelines for hydrogen: Legal and policy considerations. *Energy Reports, 8*(16), 815–820. https://doi.org/10.1016/j.egyr.2022.11.063

Jorschick, H., Preuster, P., Dürr, S., Seidel, A., Müller, K., Bösmann, A., & Wasserscheid, P. (2017). Hydrogen storage using a hot pressure swing reactor. *Energy & Environmental Science, 10*(7), 1652–1659. https://doi.org/10.1039/C7EE00476A

Kapoor, R., & Klueter, T. (2021). Unbundling and managing uncertainty surrounding emerging technologies. *Strategy Science, 6*(1), 62–74. https://doi.org/10.1287/stsc.2020.0118

Kaul, A., Hagedorn, M., Hartmann, S., Heilert, D., Harter, C., Olschewski, I., Eckstein, L., Baum, M., Henzelmann, T., & Schlick, T. (2019). Automobile Wertschöpfung 2030/2050: Studie im Auftrag des Bundesministeriums für Wirtschaft und Energie. IPE Institut für Politikevaluation GmbH; fka GmbH; Roland Berger GmbH. Retrieved June 26, 2023, from https://www.bmwk.de/Redaktion/DE/Publikationen/Studien/automobile-wertschoepfung-2030-2050.pdf?__blob=publicationFile&v=16

KTH, & QiE, P. (2019). Detailed LCA/LCCA assessment of environment and cost impact of E-roads (Road Infrastructure Impact & Solutions). FABRIC. http://www.fabric-project.eu/www.fabric-project.eu/images/Deliverables/FABRIC_D53_4-LCA_eRoads_Final.pdf

Loisel, R., Pasaoglu, G., & Thiel, C. (2014). Large-scale deployment of electric vehicles in Germany by 2030: An analysis of grid-to-vehicle and vehicle-to-grid concepts. *Energy Policy, 65*, 432–443. https://doi.org/10.1016/j.enpol.2013.10.029

Parizet, E., Janssens, K., Poveda-Martinez, P., Pereira, A., Lorenski, J., & Ramis-Soriano, J. (2016). NVH analysis techniques for design and optimization of hybrid and electric vehicles. In N. Campillo-Davo & A. Rassili (Eds.), *NVH analysis techniques for design and optimization of hybrid and electric vehicles* (pp. 313–355). Shaker Verlag.

Preuster, P., Alekseev, A., & Wasserscheid, P. (2017). Hydrogen storage technologies for future energy systems. *Annual Review of Chemical and Biomolecular Engineering, 8*, 445–471. https://doi.org/10.1146/annurev-chembioeng-060816-101334

Preuster, P., Papp, C., & Wasserscheid, P. (2017). Liquid organic hydrogen carriers (LOHCs): Toward a hydrogen-free hydrogen economy. *Accounts of Chemical Research, 50*(1), 74–85. https://doi.org/10.1021/acs.accounts.6b00474

Reitz, R. D., Ogawa, H., Payri, R., Fansler, T., Kokjohn, S., Moriyoshi, Y., Agarwal, A. K., Arcoumanis, D., Assanis, D., Bae, C., Boulouchos, K., Canakci, M., Curran, S., Denbratt, I., Gavaises, M., Guenthner, M., Hasse, C., Huang, Z., Ishiyama, T., et al. (2020). IJER editorial: The future of the internal combustion engine. *International Journal of Engine Research, 21*(1), 3–10. https://doi.org/10.1177/1468087419877990

Rüde, T., Dürr, S., Preuster, P., Wolf, M., & Wasserscheid, P. (2022). Benzyltoluene/perhydro benzyltoluene – Pushing the performance limits of pure hydrocarbon liquid organic hydrogen carrier (LOHC) systems. *Sustainable Energy & Fuels, 6*(6), 1541–1553. https://doi.org/10.1039/D1SE01767E

Sartbaeva, A., Kuznetsov, V. L., Wells, S. A., & Edwards, P. P. (2008). Hydrogen nexus in a sustainable energy future. *Energy & Environmental Science, 1*(1), 79. https://doi.org/10.1039/b810104n

Schaeffler Technologies AG (Ed.) (2014). *E-Wheel Drive*. Retrieved July 09, 2023, from https://www.schaeffler.de/de/news_medien/mediathek/downloadcenter-detail-page.jsp?id=37159745

Schaeffler Technologies AG (Ed.) (2023). Technologien für hybride und rein elektrische Antriebssysteme. Retrieved July 09, 2023, from https://www.schaeffler.de/de/produkte-und-loesungen/automotive-oem/hybride_elektrische_antriebssysteme/

Schüth, F., Palkovits, R., Schlögl, R., & Su, D. S. (2012). Ammonia as a possible element in an energy infrastructure: Catalysts for ammonia decomposition. *Energy & Environmental Science, 5*(4), 6278–6289. https://doi.org/10.1039/C2EE02865D

Schwarzer, C. (2019, March 21). Bremsenergierückgewinnung und ihr Wirkungsgrad: Wer bremst, gewinnt. heise online. Retrieved June 26, 2023, from https://www.heise.de/autos/artikel/Bremsenergierueckgewinnung-und-ihr-Wirkungsgrad-4340576.html

Specht, M. (2020, January 23). Elektroautos sind die besseren Gebrauchtwagen: Weniger Verschleiß, längere Lebensdauer. Der Spiegel. Retrieved June 26, 2023, from https://www.spiegel.de/auto/fahrkultur/elektroautos-sind-die-besseren-gebrauchtwagen-weniger-verschleiss-laengere-lebensdauer-a-89f4d920-ed88-4c9e-9747-310f685f418a

Tschöke, H., Gutzmer, P., & Pfund, T. (2019). *Elektrifizierung des Antriebsstrangs: Grundlagen - vom Mikro-Hybrid zum vollelektrischen Antrieb*. Springer Vieweg Berlin Heidelberg. https://doi.org/10.1007/978-3-662-60356-7

Umweltbundesamt. (2022, March 16). Marktdaten: Mobilität. Retrieved June 26, 2023, from https://www.umweltbundesamt.de/daten/private-haushalte-konsum/konsum-produkte/gruene-produkte-marktzahlen/marktdaten-bereich-mobilitaet#carsharing

U.S. Department of Energy (Ed.) (2023). *Hydrogen Pipelines. Hydrogen and fuel cell technologies office*. Retrieved, July 09, 2023, from https://www.energy.gov/eere/fuelcells/hydrogen-pipelines

van Basshuysen, R., & Schäfer, F. (Eds.). (2017). *Handbuch Verbrennungsmotor: Grundlagen, Komponenten, Systeme, Perspektiven* (8th ed.). Springer Vieweg Wiesbaden. https://doi.org/10.1007/978-3-658-10902-8

Verband der Automobilindustrie e. V. (2018). WLTP – neues Testverfahren weltweit am Start: Fragen und Antworten zur Umstellung von NEFZ auf WLTP. Retrieved June 26, 2023, from https://www.vdik.de/wp-content/uploads/2019/09/WLTP_Fragen-und-Antworten-zum-neuen-Testverfahren.pdf

Wang, E. (2021). *The potential of electric vehicles*. Princeton University. Retrieved June 26, 2023, from https://psci.princeton.edu/tips/2021/2/20/the-potential-of-electric-vehicles

Weigelt, M. (2022). Multidimensionale Optionenanalyse alternativer Antriebskonzepte für die individuelle Langstreckenmobilität. In *FAU Studien aus dem Maschinenbau* (Vol. Band 412). FAU University Press. https://doi.org/10.25593/978-3-96147-608-4

Zapf, M., Pengg, H., Bütler, T., Bach, C., & Weindl, C. (2021). *Kosteneffiziente und nachhaltige Automobile: Bewertung der realen Klimabelastung und der Gesamtkosten - Heute und in Zukunft*. Springer Vieweg Wiesbaden. https://doi.org/10.1007/978-3-658-33251-8

ZF Friedrichshafen AG (2017, June 22). *Für Achshybride oder reine E-Fahrzeuge: ZF integriert leistungsstarken elektrischen Antrieb direkt in innovative Hinterachse* [Press Release]. Retrieved July 09, 2023, from https://press.zf.com/press/de/releases/release_2782.html

ZF Friedrichshafen AG (Ed.) (2023). CeTrax – Elektrischer Zentralantrieb. Systemkompetenz für elektrische Sonderfahrzeuge. Retrieved July 09, 2023, from https://www.zf.com/products/de/special_vehicles/products_64419.html

Open Access This chapter is licensed under the terms of the Creative Commons Attribution 4.0 International License (http://creativecommons.org/licenses/by/4.0/), which permits use, sharing, adaptation, distribution and reproduction in any medium or format, as long as you give appropriate credit to the original author(s) and the source, provide a link to the Creative Commons license and indicate if changes were made.

The images or other third party material in this chapter are included in the chapter's Creative Commons license, unless indicated otherwise in a credit line to the material. If material is not included in the chapter's Creative Commons license and your intended use is not permitted by statutory regulation or exceeds the permitted use, you will need to obtain permission directly from the copyright holder.

9

The Road to Net Zero and Beyond
Looking Back, Taking Stock, and Moving Forward

Markus Beckmann and Irene Feige

9.1 Reflecting on Collaborative Learning

What are the strategic pathways for sustainability-driven business transformation? With this question in mind, Bayerische Motoren Werke AG (BMW) and FAU came together in a series of expert dialogues to discuss critical topics on the Road to Net Zero. This book not only documents these dialogues but also uses them as the core for seven chapters. Each chapter explores some of the many complexities and intriguing issues involved in transforming our economy toward sustainability. Supplemented by a brief introduction to the topic and selected future research avenues, the idea was to allow readers to dive directly into the facets of the "Road to Net Zero" that interest them.

Each chapter therefore stands on its own. However, no chapter stands in isolation. On the contrary, the various steps on the Road to Net Zero all interact, are highly interdependent, and require an integrated perspective. The purpose of this concluding chapter is to reflect on this bigger picture. We will do this in four steps. First, we look back by briefly reviewing the lines of argument developed in each chapter and how they relate to each other. Second, we take stock by identifying recurring themes, critical insights, and lessons (to

---

M. Beckmann (✉)
FAU Erlangen-Nürnberg, Nuremberg, Germany
e-mail: markus.beckmann@fau.de

I. Feige
BMW AG, Munich, Germany

be) learned that we have identified across individual chapters. Third, focusing on journeying forward, we look beyond the Road to Net Zero outlined in this book and identify future questions for sustainability-driven business transformation. Finally, we reflect on the nature of these challenges and discuss the role of industry–university partnerships in addressing them.

9.2 A Summary of this Book's Storyline

The title of this book, "The Road to Net Zero," signifies an ambitious objective—achieving net-zero emissions to curb global warming. This bold ambition has profound implications. We cannot reach carbon neutrality merely by making incremental improvements within the existing fossil fuel–based economy. Instead, it necessitates a complete transformation in our business practices, with wide-ranging consequences and contributions required across different domains, as discussed in each chapter of this book.

9.2.1 Chapter 2: Setting the Course for Net Zero

Chap. 2 initiated the discussion by dissecting the scientific and political aspects of the net zero concept. It elaborated on why the Paris Agreement was a pivotal moment in global climate policy. Not only was it the first global pact supported by all major carbon emitters, including the United States, China, and India, but it also established absolute global warming temperature targets. Based on climate science, these politically agreed-upon targets translate into a defined limit to the remaining carbon budget that humanity can afford, contrasting with the relative reduction approach of the Kyoto Protocol. Consequently, the Road to Net Zero demands a radical transformation of our systems, rather than continuous enhancements to existing fossil fuel technologies and business models.

The chapter delved further into how governments can catalyze this transformative change. The expert conversation within it explored the role of government policy in promoting conditions conducive to decarbonizing the economy. This includes strategies such as pricing carbon via emission certificates and carbon taxes, enforcing suitable market regulations, and promoting infrastructure development necessary for the introduction and scaling of alternative technologies. Chapter 2 also elucidated how corporations can align their business objectives with the global Road to Net Zero. It introduces the idea of setting Science-Based Targets, which offers a tangible and scientifically substantiated framework for businesses to align their operations with global climate change mitigation efforts.

9.2.2 Chapter 3: Crafting Corporate Sustainability Strategy

Chapter 3 shifted the focus from external, science-based discussions and political decisions, such as the Paris Agreement, to the internal strategic decision-making process within companies. It explored how these external parameters influence business operations. In the past, companies often set sustainability goals that were separate from their core business strategies, typically under the umbrella of corporate social responsibility. However, the escalating urgency of climate change necessitates a fundamental change in approach. Mature strategies no longer regard sustainability as an isolated component, but integrate it holistically.

On the Road to Net Zero, true life cycle decarbonization requires that sustainability be woven into the fabric of a company's value creation process to encompass the entire value chain rather than just the company's direct operations. Chapter 3 delved into how this integrated approach revolutionizes the entire strategic process. Building upon Chap. 2, it started with strategy formulation as the phase to set reliable, Science-Based Targets. It then moved on to strategy implementation, which necessitates a unified management approach, and concluded with strategy evaluation, which demands innovative methods of measurement and reporting, thus smoothly transitioning into Chap. 4.

9.2.3 Chapter 4: The Future of Corporate Disclosure

As the focus on a company's sustainability strategy and performance intensifies, the traditional approach of reporting solely on financial indicators falls short of meeting the diverse interests of all stakeholders. Traditional reporting, primarily designed for investors, emphasizes the company's financial performance. However, in today's context, specifically on the Road to Net Zero, there is a growing need to encompass nonfinancial, sustainability-related aspects to address the information requirements of a wider array of stakeholders, including employees, governments, and society at large.

The shift toward nonfinancial or sustainability reporting has seen a significant evolution, moving from voluntary standards with limited comparability to stringent regulatory requirements advocating for enhanced transparency. Chapter 4 delved into this transition and its relation to integrated reporting. It reflected upon recent legislative changes, explored the challenges associated with measuring and selecting both financial and nonfinancial key performance indicators (KPIs), and discussed the delicate task of balancing the diverse interests of different stakeholders.

9.2.4 Chapter 5: Creating Sustainable Products

In the automotive industry, one reason why the Road to Net Zero depends on trustworthy information and comprehensive reporting is that the advent of electric vehicles has shifted the majority of life cycle emissions from the usage phase to the production phase. As a result, circular value chains and their various elements have become paramount in the operational transformation to net zero. With this in mind, Chap. 5 shifted the focus to the importance of product design. Because decisions made in product design have diverse impacts on material sourcing, manufacturing, a product's use phase, and the options for closing material flows at the end of its life, the transition to a circular economy necessitates a reimagined approach to product development. Chapter 5 therefore explored design for recycling, the substitution of scarce resources with secondary materials, and the introduction of natural materials, with a particular focus on interior design. Implementing circular design also requires a shift in service design. Services that facilitate circularity, such as product take-back programs, become essential. Alternatively, a transition in ownership from manufacturers to service providers, emphasizing access and performance-based business models, can help execute circular strategies. Consequently, product design and service design must evolve in tandem.

9.2.5 Chapter 6: Transforming Value Chains for Sustainability

The conversation about sustainability in product development inevitably leads to the issue of procuring scarce and valuable resources. On the Road to Net Zero, the substitution of primary materials with secondary materials offers a crucial lever for reducing carbon emissions over the entire life cycle. Chapter 6, therefore, deepened the material flow analysis by zooming into this potential, as well as the challenges of sourcing scarce and valuable resources for electric mobility. Manufacturing batteries and electric drivetrains, in particular, requires energy-intensive materials that are not only limited in quantity, but are also concentrated in a handful of countries worldwide. In addition to its relevance to greenhouse gas (GHG) emissions, this constraint underscores the call for closed-loop supply chains that incorporate secondary materials into production.

However, realizing the vast potential of these material flow loops to achieve carbon reduction and environmental impact goals is not without hurdles. Substantial technological, organizational, and regulatory obstacles exist at each stage of the circular value chain. Because circularity requires robust data, further digital advances, such as the digital battery passport and appropriate data-sharing methods, are essential. In addition, as later noted in Chap. 8 and its discussion of the role of technology, Chap. 6 provided important background on why continued innovation in battery storage technologies can significantly contribute to offsetting existing resource scarcity and will strongly shape the blueprint of future sustainable supply chains.

9.2.6 Chapter 7: Sustainability in Manufacturing

Before Chap. 8 delved into this role of technology, Chap. 7 focused on the transformation of traditional factories into green, sustainable ones as a pivotal element in the operational transition to net zero. Over the last two decades, production optimization has been a focal point for researchers and practitioners. While the previous shift toward operational excellence was primarily internal, an integrated strategy now necessitates a broader, system-wide consideration, which ties back to Chaps. 3, 5, and 6. To illustrate this shift, Chap. 7 used the BMW iFACTORY as an example, which combines lean systems, digital technologies, and circular production to address sustainability comprehensively. Chapter 7 thus highlighted that the responsibility for implementing decarbonization and circularity begins, but does not end, within a company's own operations. It also emphasized that sustainable production is not a one-off project, but a continuous process that builds on a company's expertise in quality or lean management. This type of continuous learning requires access to multiple sources of data, from production machines to manual processes, again underscoring the relevance of data and transparency.

9.2.7 Chapter 8: The Power of Technological Innovation

The disruptive transformation required for the Road to Net Zero is not possible based (solely) on the basis of incremental efficiency improvement in our current fossil fuel–based technologies. This is particularly true in the mobility sector. Chapter 8 examined technology alternatives that can replace the internal combustion engine (ICE) as the dominant drivetrain technology. The chapter conducted a systematic analysis of future drivetrains, from electric drivetrains to synfuel internal combustion engines (ICEs) and fuel cells, discussing the eco-efficiency as well as the challenges and opportunities of each technology and outlining potential future technological developments.

A key finding of the chapter was that there is no technological silver bullet. Instead, each technology will be part of the decarbonization roadmap dictated by specific use cases and contexts. Battery electric vehicles are likely to dominate personal transport, while hydrogen could complement in areas such as long-distance and heavy-duty transport or in sparsely populated contexts. Regardless of the underlying green technology, future vehicles will be Autonomous, Connected, Electric, and Shared (ACES), with each aspect contributing to sustainability when effectively integrated. However, zero-emission technologies need an enabling energy ecosystem with sufficient renewable energy and innovative charging infrastructure. This is a task to be addressed not only by private companies but also by public policies that promote market regulation, incentives, and investments for this enabling environment. Chapter 8 thus closed the loop by linking back to the first expert conversation and the public policy discussion in Chap. 2.

9.2.8 Looking Back on the Road to Net Zero

Upon reflecting on the summary presented above, we would like to share two crucial observations. First, the expert discussions documented in this book and the seven resulting chapters do not purport to offer a comprehensive view of all the challenges and questions surrounding the Road to Net Zero. Rather, while addressing many vital topics, some important perspectives are absent. For instance, the role of consumers is touched upon (e.g., in Chap. 3 as a sustainability driver or in Chap. 5 when discussing service design to promote circular consumer behavior) but not explored in depth, and the interaction between automotive mobility and broader mobility systems is mentioned (e.g., in Chap. 7 on smart cities) but not analyzed thoroughly. We anticipate future discussions to tackle these and other underexplored aspects.

Second, the summary outlines the book's chapters in a linear fashion, implying a sequential progression. However, the real-world transition to net zero is far from linear, as all elements must occur simultaneously, with numerous interconnections and feedback loops arising between the various topics. Thus, the following section identifies and discusses recurring themes that link the different chapters and emerging insights as potential lessons to be learned across chapters.

9.3 Insights and Themes across Chapters

As editors, we had the rewarding experience of reviewing all the chapters multiple times, often oscillating between different sections. This iterative process illuminated recurring themes and insights that echo throughout the various chapters and, at times, across the entire book. We present some of these observations in this section. It is important to note that not all these insights are novel or counterintuitive. To those of us deeply immersed in sustainability management, some findings may seem evident and familiar. However, their recurrence does not diminish their importance but underlines their significance. We also share these insights with readers in mind who may be new to the discourse on sustainability management. Regardless of whether these insights seem novel to you, resonate with your experiences, or provoke critical thought, our aim is to stimulate a thoughtful discourse.

9.3.1 Sustainability Demands Both Integrative Thinking and Integrative Management

Sustainability is widely recognized to necessitate integrative thinking due to the multifaceted nature of sustainability challenges. These challenges cut across various disciplines, sectors, and stakeholder interests, requiring an understanding that goes beyond the siloed knowledge of individual disciplines. Integrative thinking enables the synthesis of diverse perspectives and the ability to see interconnections and interdependencies. This is critical for devising holistic, effective strategies for sustainability that consider the environmental, social, and economic dimensions in tandem rather than in isolation.

In this book, experts from different disciplines exemplify this integrative thinking with a focus on integrative strategy (Chap. 3), integrative reporting (Chap. 4), the integration of sustainability in product development (Chap.

5), value chains (Chap. 6), or the factory of the future (Chap. 7). These contributions highlight that integrative thinking is more than the simultaneous consideration of ecological, social, and economic factors. It also requires a holistic view when embracing a life cycle perspective, integrating sustainability aspects from raw material extraction to a product's end of life.

However, integrative thinking does not come naturally. While functional differentiation and specialization across disciplines are essential for gaining the detailed understanding required to address specific challenges on the Road to Net Zero, they also create fragmentation that requires reintegration. This leads to a strong call for an integrative management approach, both within and between organizations.

Integrative management emphasizes collaboration and coordination across various units within an organization and among diverse external stakeholders. In the context of sustainability, this might mean aligning diverse internal functions—some of them represented in this book—on sustainability goals. It could also involve collaboration with external stakeholders, such as suppliers, customers, governments, nongovernmental organizations (NGOs), and communities, to cocreate sustainable value.

So what is the takeaway? The integrative thinking required for sustainability must be complemented by appropriate forms of integrated management. This does not only mean incorporating sustainability into individual corporate functions; it also includes creating organizational structures that promote cross-functional collaboration and knowledge sharing. Furthermore, it is essential to align incentives that bind individual roles and responsibilities with the bigger sustainability picture.

9.3.2 Sustainability Is a Moving Target

That "sustainability is a moving target" is not only a well-known adage in our field, it is also a practical reality that we experienced while working on this book. From the expert discussions in 2021 to the completion of this book, we noted a myriad of changes in our conversations. Some details that were current during our initial discussions may appear outdated now. Without changing their substance, we have carefully adjusted these sections where appropriate. We have also encapsulated the expert dialogues within comprehensive chapters that focus on the emerging longer-term picture, with the goal of providing enduring value to our readers.

Sustainability is a moving target primarily because it operates within a complex, dynamic system characterized by a continual change. Social, economic, environmental, and technological factors are all in a constant state of flux, and each of these influences our understanding of sustainability, as well as the drivers and means for addressing it. For instance, as scientific knowledge about climate change deepens, our goals for carbon reduction may become more aggressive (Chap. 2). Similarly, as market demands and regulations change, so do the drivers for sustainability strategy (Chap. 3). Since the start of this book project, changing regulatory requirements have also massively reshaped the field of sustainability reporting (Chap. 4). Likewise, geopolitical disruptions have changed the discourse on sustainable and resilient value chains between the initial expert conversations (2021) and this book's publication (2023) (Chap. 6). Finally, rapid technological change, including accelerated advances in artificial intelligence, offers new opportunities to promote circularity (Chap. 5), sustainable manufacturing (Chap. 7), and, of course, the race for green drivetrains (Chap. 8).

So what is the takeaway? On the Road to Net Zero, we must continually refresh our understanding of our destination and the path to reach it. The multifaceted nature of sustainability means that advancements in one area could trigger new challenges and opportunities in others. Therefore, sustainability is not a fixed target, but an ongoing, evolving journey that demands constant reassessment and adjustment of our strategies and goals. This journey requires iterative learning and, crucially, unlearning. We must reevaluate and may need to discard yesterday's answers and practices and develop new ones tomorrow. It necessitates questioning established responsibilities, business models, technologies, and the notion of going it alone. Unlearning is challenging, especially in isolation. Hence, collaboration and exposure to alternative perspectives are vital for learning and prospering, which leads us to our next insight.

9.3.3 Sustainability Is a Race You Cannot Win Alone

Sustainability challenges, such as climate change, are systemic problems that require systemic solutions. The Road to Net Zero is, therefore, about systemic change. However, as nearly every expert conversation highlighted, no company or organization can single-handedly achieve the systemic changes needed to transition to a sustainable future. Instead, collaboration is needed to secure and pool resources, share knowledge, and align efforts around common goals.

Collaboration with diverse stakeholders is nothing new for companies. In fact, cocreating value with customers, suppliers, investors, employees, and communities is at the core of what defines a well-managed firm and sustainable growth. However, the Road to Net Zero calls for deeper and more nuanced forms of partnerships and collaboration. With this book's focus on the automotive industry in mind, we would like to highlight three of them.

First, public–private collaboration is essential for setting the stage for sustainable mobility solutions. This includes not just relevant energy market regulation and carbon pricing, but also the establishment of necessary infrastructure for electric or hydrogen mobility, from renewable energy generation to distribution, storage, and charging infrastructure (Chaps. 2 and 8). Second, on the Road to Net Zero, companies are being held accountable for the life cycle impact of their products. This is especially pertinent for the automotive industry, where vehicle electrification shifts emissions from usage to the upstream value chain (Chap. 6). To truly take responsibility for their full life cycle impact, companies must forge more profound collaborations along their value chains. This involves exchanging sustainability-related data, codeveloping and integrating greener technologies, and establishing circular material flows (Chap. 5). For instance, automotive OEMs can significantly reduce their carbon footprints by replacing primary materials with secondary ones. However, these solutions demand more sophisticated collaboration along the value chain, as Chap. 6 demonstrated for electric batteries. Third, decarbonizing entire industries and sectors requires competitors to collaborate to establish robust industry standards and an equitable playing field for sustainability (Chap. 3). This is already happening in the automotive industry with initiatives like Drive Sustainability, a partnership of leading OEMs improving supply chain sustainability, and the emerging data ecosystem Catena-X (Chaps. 5 and 6).

So what is the takeaway? Collaboration is undoubtedly a buzzword, and it is hard to argue against it. However, while it sounds simple, in reality, it can be a complicated dance. Coordinating different partners amplifies the complexity and necessitates the reconciliation of divergent, often conflicting, interests. While the Road to Net Zero is grounded in a shared commitment to a sustainable future, the perspectives and motivations of companies, regulators, and civil society actors often diverge. Even among themselves, OEMs and suppliers vie for value, and competitors seek to outdo each other. Even within industries, companies compete for a slice of the pie. Competition and diverging interests do not disappear in the pursuit of sustainability; instead, it is about forming partnerships that respect these differences and align them toward a common goal. Hence, companies must hone their partnership skills, including the ability to compete within a set framework while jointly crafting a better one.

9.3.4 It Is All About Data: Measurable Indicators, Targets, Transparency, and Digitalization

Data, along with measurable indicators, targets, transparency, and digitalization, are at the heart of the Road to Net Zero. The book's opening chapter set the tone with the principle, "what gets measured, gets done," emphasizing that the journey toward sustainability must be grounded in reliable data and evidence, especially when it comes to the decarbonization of industry. Without them, efforts lack direction and tangibility.

Defining the right indicators is a crucial first step. For climate change, this seems straightforward. Here, carbon dioxide (CO_2) and the other greenhouse gases identified in the Kyoto Protocol and translated into CO_2 equivalents form the basis for the global climate policy discourse (Chap. 2). These indicators, in turn, enable effective target setting, both for the global Road to Net Zero (Chap. 2) and for corporate strategy (Chap. 3). At the global level, the 2 °C and 1.5 °C global warming goals and the implied remaining carbon budget present benchmarks against which global climate action can be measured, revealing our current dramatic shortcomings. Building on the CO_2 indicator to translate global targets to the corporate level, companies can leverage frameworks like Science-Based Targets to align their decarbonization efforts with the decarbonization paths needed at the planetary level (Chap. 2).

Along with the indicators and targets, the right measurement scope is equally important. While at a planetary level, this is straightforward (all of humanity's emissions are included), the picture is more complex at the corporate level. Here, some emissions are caused by a company's own operations, while others occur upstream or downstream in the value chain (Chaps. 3, 5, 6, and 7). A life cycle approach to measuring emissions, therefore, accounts for CO_2 emissions from all stages of a product's life—from the extraction of raw materials to the disposal of the product. This comprehensive view reveals hidden emissions and enables better decision-making. At the same time, it adds complexity, as Scope 3 emissions (occurring in the value chain) are much more difficult to influence and measure.

However, having the right indicators, targets, and scope is pointless without access to high-quality data. Particularly when it comes to driving decarbonization in a complex system like a value chain, having precise, timely, and comprehensive data is paramount. This type of data allows businesses to identify hotspots for improvement, implement changes, and track their success. However, generating high-quality data is easier said than done. Currently, the majority of Scope 3 emissions are estimated using databases. Actual, real-time data are scarce. In addition, suppliers are reluctant to share data. Overcoming these challenges requires collaboration and technological innovation, with digital platforms like Catena-X demonstrating how digital solutions can facilitate data sharing on the Road to Net Zero.

The role of digital solutions in accelerating progress on the Road to Net Zero cannot be overstated. Digitization has the potential to significantly enhance the transparency of carbon emissions in value chains (Chap. 6). Technologies such as blockchains, artificial intelligence (AI), and the Internet of things (IoT) can provide unprecedented visibility into the environmental footprint of products and services throughout their life cycles (Chap. 5). For example, IoT devices can capture real-time emissions data at every stage of production (Chap. 7) and distribution, from raw material extraction to end-user consumption. AI can subsequently process this enormous amount of data to generate actionable insights, identify emissions hotspots, and propose effective mitigation strategies. Blockchain technology enables securing this data, ensuring its integrity, and making it tamper-proof. This shared, decentralized ledger allows every participant in the value chain to access and verify the same emissions data, fostering a culture of accountability and collaboration. Thus, by harnessing the power of these digital technologies, we can create a data-driven, transparent, and trustworthy system for tracking and reducing carbon emissions across value chains.

Finally, data are important for internal decision-making. Reporting this type of data, with all its complexities and nuances, is crucial to ensuring transparency and building trust among stakeholders (Chap. 4). Transparency promotes accountability, while also fostering an environment in which best practices are shared and replicated and accelerate the industry-wide transition to sustainability.

So, what is the takeaway? Progress on the Road to Net Zero depends on the quality of the data that guides us. This means that companies should take a science-based approach to defining reliable metrics and setting targets that are aligned with societal sustainability goals. To consider holistic life cycle impacts, collaboration across the value chain requires the flow of data and information along it. Digitalization provides the game-changing innovations needed to generate, exchange, process, and disclose data at this new level of complexity.

9.3.5 Not Everything That Matters Can Be Measured (Accurately)

While data, metrics, and measurement are crucial on the Road to Net Zero, many chapters emphasized that factors such as vision, leadership, and culture are equally essential to a successful transition to sustainability (Chaps. 1, 3, 5, 6, and 7).

Metrics are undeniably important for tracking progress and facilitating decision-making, but they are always selective and therefore have their limitations. One of the key challenges lies in the multidimensionality of sustainability. Sustainability encompasses a wide array of elements, from environmental protection to social equity and economic viability. While some of these aspects can be quantified and measured, others are qualitative and intangible, making it hard to aggregate them into a single, comprehensive sustainability index.

Moreover, the sustainability transition discussed in this book is taking place in a 'VUCA' world (Chap. 3) characterized by volatility, uncertainty, complexity, and ambiguity. Predicting future scenarios or outcomes involves a multitude of variables and assumptions, many of which are subject to change due to unpredictable factors. Therefore, while metrics can guide our actions and decisions, they cannot perfectly predict the future or fully capture the complexity of sustainability.

This is where the importance of vision, leadership, culture, and long-term commitment comes into play (Chaps. 5, 6, and 7). The role of leadership is not just to understand and use metrics but to provide a vision of the future that transcends these measurements. Leaders need to inspire and motivate and to foster a culture of sustainability that is rooted in values and beliefs as much as in data and evidence. This culture, in turn, can shape behaviors and decisions in ways that cannot be precisely measured but are nonetheless critical to achieving sustainability goals.

A good illustration of this is the 1.5 °C global warming target that underpins the Road to Net Zero. While this target is informed by scientific data on the potential impacts of climate change, the decision to set this particular target was ultimately a political one (Chap. 2). It represents a collective vision of a future in which we limit global warming to a level that, according to scientific consensus, could prevent the most catastrophic effects of climate change. This vision and the leadership required to pursue it are essential complements to the metrics that guide our path toward sustainability.

So what is the takeaway? The transition to sustainability requires a balanced approach, combining both technological advancements and evidence-based decision-making with shared values and a clear vision. Metrics alone, without a guiding vision, can lack direction and may not lead to meaningful change. Conversely, a vision without factual evidence might result in superficial or ineffective solutions. Achieving sustainability demands the thoughtful integration of both elements, creating a comprehensive strategy that is firmly rooted in scientific facts and guided by shared values and aspirations.

9.4 Beyond the Road to Net Zero

This book has brought together selected perspectives that have their role to play on the Road to Net Zero. In doing so, the primary focus was on the urgent issue of climate change. However, as we underscored earlier, sustainability is a moving target that is constantly evolving in response to our growing understanding of our relationship with the planet and the realization of our collective responsibilities. Therefore, reaching net zero is just a milestone, not the final destination. In this section, we highlight a few exemplary topics that merit further exploration but are beyond the scope of this book. Although these topics have been touched upon within this book, they invite a more comprehensive exploration in our ongoing quest for sustainability. The list is far from complete but suffices to show the potential for further collaboration between universities and industry.

9.4.1 Beyond Decarbonization

This book has primarily focused on the Road to Net Zero, acknowledging the urgent need to reduce CO_2 emissions in light of the ongoing climate crisis. This is a significant step forward in addressing the sustainability challenge. However, we must understand that it is not the only pressing environmental issue we face today.

Beyond carbon emissions, other pressing ecological challenges exist, such as the significant footprint of our material resource consumption. This footprint affects not only carbon emissions but also biodiversity, our water resources, and social issues, including human health. The Global Resources Outlook by the United Nations Environment Programme (UNEP) in 2019 revealed that resource extraction has tripled since 1970, even though the population has only doubled during the same period. Extractive industries contribute to half of the world's carbon emissions and account for more than 80% of biodiversity loss.

These consumption patterns, with the world utilizing over 92 billion tons of materials annually and growing at a rate of 3.2% per year, are not sustainable. Agricultural land-use changes account for over 80% of biodiversity loss and 85% of water stress, with the extraction and initial processing of metals and minerals accounting for a significant proportion of health impacts from air pollution and global carbon emissions.

To address these issues, we need to decouple economic growth from material consumption. This can be achieved through a circular economy, which not only reduces CO_2 emissions but also reduces the strain on our planet's finite resources. The importance of slowing down, narrowing, and closing material flows for this purpose has been described in detail in Chap. 5 of this book. There is already a broad knowledge base in sustainability management for implementing circularity through various strategies. However, compared with the climate-related Road to Net Zero, there is still a critical path ahead.

In fact, the significance of the Road to Net Zero goes beyond its climate change implications. It symbolizes humanity's first attempt to translate critical planetary boundaries into policy frameworks and science-based corporate targets. The "Road to Net Zero" recognizes that human prosperity and economic growth must be aligned with the ecological carrying capacity of our planet, with the climate system being one such planetary boundary. The Paris Agreement and its related frameworks provide an accepted benchmark for climate change. However, for other planetary boundaries, such as biodiversity, the development of adequate indicators, global targets, development pathways to meet them, company-specific targets, and standards for measuring and comparing performance is much more in its infancy.

In this book, we have focused on the Road to Net Zero greenhouse gas emissions. However, humanity must also bend the curve regarding the loss of biodiversity, soil, or freshwater reserves. With the Global Biodiversity Framework agreed upon in December 2022 at the United Nations' Biodiversity Conference COP15 in Montreal, a New Road to Net Zero could emerge regarding biodiversity loss. A business-related initiative linked with this discourse is the Science-Based Targets for Nature (SBTN), which extends the successful Science-Based Targets Initiative (SBTI) for climate to other aspects of nature. SBTN seeks to develop quantifiable, scientifically backed objectives to mitigate the impacts of environmental degradation and biodiversity loss. The initiative's success remains to be seen, but its potential to influence corporate activities is encouraging.

Alongside ecological challenges, businesses are also grappling with their broader social impact, including their duty to respect and uphold human rights. As industries transition toward electrification, green energy, and circularity, this affects the complexity of global supply chains, presenting new risks of indirect involvement in human rights abuses. Corporations are also tasked with ensuring fair labor practices, not only within their own organizations but also within their supply chains, which span multiple jurisdictions with varying labor standards. The rapid advancement of technology, such as automation and artificial intelligence, has introduced new challenges in preserving privacy and preventing discrimination.

As on the Road to Net Zero, businesses must foster more effective collaborations with governments, NGOs, and communities to successfully address these issues. In addition, they must establish robust grievance mechanisms and remediation processes to respond effectively when things go awry. The importance of transparency and accountability cannot be overstated in these matters. However, just as discussed for environmental challenges, the development of comprehensive indicators and standards for measuring, comparing, and reporting human rights performance remains an ongoing process, again reflecting the nature of sustainability as a moving target.

Finally, grasping the interconnectedness and intersectionality of various sustainability challenges is crucial. For instance, while the electrification of mobility presents a promising path toward lowering CO_2 emissions, the increasing demand for certain minerals can inadvertently exacerbate biodiversity loss and result in human rights challenges. Similarly, the shift toward a circular economy is not devoid of complexity. This might necessitate managing trade-offs between reducing GHG emissions, moderating water usage, minimizing biodiversity impacts, optimizing required land mass, and limiting hazardous emissions. In this context, businesses transitioning to circularity must develop a nuanced understanding of these trade-offs and establish rigorous criteria for evaluating and comparing them when making pivotal decisions. As we continue our journey toward a sustainable future, it is essential that our approach is comprehensive, balanced, and cognizant of these interwoven facets of sustainability.

9.4.2 Beyond Reducing Negative Impacts

As our discussion of the Road to Net Zero focused on how to sharply diminish carbon emissions, the spotlight has invariably rested on curtailing a company's adverse impact on its environment. Unarguably, mitigating harm carries immense significance—a truth we expounded in our preceding discourse. However, an overemphasis on the reduction of harm can cast a shadow on another equally important aspect—the positive impacts that society expects companies to create. Traditionally, these positive contributions encompass a wide range of factors, from the utility of their products and services in addressing various human needs and the financial and personal development opportunities offered to employees to the financial returns disbursed to investors and society via taxes.

As we traverse the Road to Net Zero, acknowledging the positive roles of corporations becomes crucial to ensure that well-meaning attempts to pare down negative effects, such as decreasing carbon emissions, do not inadvertently cause unproportionly harm elsewhere by undermining the value companies bring to their stakeholders.

Understanding how companies can create positive impacts is also crucial to achieving net zero goals, as discussed in this book. The term "net zero" itself suggests a balance—it does not denote the absolute absence of emissions; rather, it suggests the idea that remaining emissions added to the atmosphere can be offset by emissions eliminated or sequestered elsewhere.

In the Paris Agreement, the net zero goal mandates that, by 2050, global emissions must be as close to zero as feasible, with any lingering emissions reabsorbed from the atmosphere by oceans and forests, for instance. This emphasizes the need for carbon removal, hence shifting the focus toward measures with a positive impact. Similarly, the SBTI Net-Zero Standard states that companies wishing to adhere to this ambitious standard must not only establish robust reduction targets (at least 90% emission reduction by 2050) but also neutralize any remaining emissions through permanent carbon removal and storage.

This focus on activities with a positive impact extends beyond just net zero carbon emissions to include net zero targets for biodiversity and water. Here, for any biodiversity loss that cannot be avoided, businesses need to neutralize these remaining negative effects with positive impact measures. Positive impact activities are therefore pivotal in achieving any net zero goal. However, even upon reaching the net zero landmark, net zero emissions or biodiversity losses only signal a cessation of future harm to our climate and natural environment. The existing damages and depleted ecosystems remain unaddressed, which is where regenerative business practices come into play.

Regenerative business practices are an emerging paradigm that aims to restore, renew, and revitalize their own sources of energy and materials. This concept takes sustainability a step further, moving beyond merely reducing harm to actively repairing and enhancing the environments and communities in which a business operates. These practices embody the essence of regeneration—they are, by definition, net positive.[1] They enable companies to design systems that not only sustain but also enhance the capacity of the environment and communities to flourish. Companies can apply such regenerative practices in their own operations, encourage their suppliers to do so, or collaborate with third partners (e.g., when restoring natural ecosystems or removing carbon from the atmosphere).

These capabilities to generate positive impacts, individually or in collaboration with others, open up thrilling prospects beyond the net zero milestone. Hence, once we have traversed the Road to Net Zero, the next stage of ambition could be to embark on the Road to Net Positive.

9.4.3 Beyond Single Trajectories That Ignore the Role of Space

Tackling climate change via the Road to Net Zero is a worldwide effort, but it is crucial to recognize that solutions and contributions are context-specific. This book has touched lightly upon this geographical aspect, but it needs more in-depth exploration in future research and cross-sector dialogue.

Place matters enormously when it comes to transformation pathways toward sustainability. From a business viewpoint, various factors, such as the enabling environment, strategic drivers, potential alternatives, practical constraints, and benefits of sustainability engagement, are all influenced by where the company operates. Here, companies often encounter fragmented and sometimes contradictory environmental contexts that pose the challenge of forming a unified, consistent strategy.

To start with, in Chaps. 2 and 8, and beyond, this book highlights the role of public policy in shaping corporate sustainability strategies, ranging from energy market and carbon pricing regulations to those concerning products and technologies (such as fleet emissions or the ban of internal combustion engines) and infrastructure development (such as charging infrastructure). However, public policies are far from uniform globally. For example, emerging regulations for recycling electric batteries vary significantly between China and the European Union, meaning that the requirements for companies' sustainability strategy differ depending on the regulatory context.

In the automotive industry, providing mobility has, by definition, a spatial perspective. It makes a difference whether mobility is provided in an urban

---

[1] Note that "positive" here refers to the desirable, positive impact, not to remaining undesirable effects that have not been reduced to zero.

context in the Netherlands or in a rural context in Brazil. The spatial context will influence customer needs, the availability of charging infrastructure, and the suitability of different technologies to meet those needs.

Furthermore, when assessing the sustainability of value chains, it is critical to consider spatial factors affecting the life cycle impacts of technologies and usages. Factors such as the availability of green energy, water scarcity, biodiversity impacts, and emissions from transport and logistics are all location-dependent. Hence, the most sustainable alternative in one location might not be the same in another. External drivers like societal expectations and customer and employee needs play a key role in shaping a company's sustainability strategy (Chap. 3). These societal drivers vary greatly between regions. Therefore, global companies must balance meeting diverse expectations while maintaining their internal consistency. Considering these spatial contexts, the journey to net zero will follow different trajectories in different regions.

This book evolved from a discussion among experts from the same spatial neighborhood. The FAU and BMW headquarters are both located in Bavaria, Germany. However, for a more comprehensive understanding, we need to foster dialogue with stakeholders from other global regions like China, the United States, and others. Future exchanges should focus more on understanding these differing contexts, their influencing factors, their impact on sustainability, and how companies, regulators, and other stakeholders can align multiple transition pathways on our shared planet.

9.5 The Future of Industry–University Partnerships

The journey toward sustainability requires innovative technologies and robust policy frameworks, as well as dynamic collaborations that transcend traditional boundaries. This book stands as a testament to this type of collaborative spirit by documenting and expanding on the dialogues of experts from BMW and FAU, who came together to discuss critical landmarks on this journey.

These dialogues underscore the potential of university–industry partnerships in addressing sustainability challenges. These challenges are complex and multifaceted, demanding an array of perspectives and interdisciplinary dialogues. Universities, particularly full-spectrum universities like FAU, are uniquely equipped to facilitate such dialogues. They encompass a wide spectrum of disciplines and maintain a crucial outside position, providing a healthy distance and independence from business, political actors, and civil society. This independence allows them to critically assess the status quo and propose innovative solutions grounded in rigorous research.

However, the dialogues in this book, while insightful, have largely represented rather homogenous views. This points to the need for a greater diversity of viewpoints in our discussions around sustainability. Controversies about sustainability should not be viewed as a hindrance, but rather as a source of innovation and learning. If orchestrated constructively, they can spark creative solutions and drive progress. In these dialogues, it is important to respect the differing identities, needs, and integrities of various actors, recognizing that each contributes a unique perspective to the overall discussion.

While this book has focused on the collaboration of two strong partners in Bavaria, BMW and FAU, future collaboration must transcend regional and institutional boundaries. Universities, with their inherently international nature and rich tapestry of research collaborations, are well suited to facilitate such partnerships. They can serve as a platform where perspectives from different regions, disciplines, and sectors can coalesce, fostering a rich discourse that can fuel sustainability transformation.

Furthermore, universities serve as a natural bridge and forum for discussion between generations. They are places where enthusiastic and sustainability-driven youth interact with experienced professionals and academics. This intergenerational dialogue can stimulate fresh thinking and drive momentum toward sustainability goals.

In conclusion, the power of university–industry partnerships in driving the transition to sustainability cannot be overstated. These partnerships offer an invaluable platform for interdisciplinary dialogues, viewpoint diversity, and international collaborations. More than that, they provide a space where respect for differing identities and needs, and the passion of different generations for sustainability, can come together to catalyze transformative action on the Road to Net Zero and beyond.

Open Access This chapter is licensed under the terms of the Creative Commons Attribution 4.0 International License (http://creativecommons.org/licenses/by/4.0/), which permits use, sharing, adaptation, distribution and reproduction in any medium or format, as long as you give appropriate credit to the original author(s) and the source, provide a link to the Creative Commons license and indicate if changes were made.

The images or other third party material in this chapter are included in the chapter's Creative Commons license, unless indicated otherwise in a credit line to the material. If material is not included in the chapter's Creative Commons license and your intended use is not permitted by statutory regulation or exceeds the permitted use, you will need to obtain permission directly from the copyright holder.

Printed by Printforce, the Netherlands